Audel
Industrial
Multi-Craft Mini-Ref

Thomas Bieber Davis

WILEY

JOHN WILEY & SONS, INC.

Copyright © 2012 by John Wiley & Sons, Inc. All rights reserved

Published by John Wiley & Sons, Inc., Hoboken, New Jersey

Published simultaneously in Canada

For general information about our other products and services, please contact our Customer Care Department within the United States at (800) 762-2974, outside the United States at (317) 572-3993 or fax (317) 572-4002.

Wiley publishes in a variety of print and electronic formats and by print-on-demand. Some material included with standard print versions of this book may not be included in e-books or in print-on-demand. If this book refers to media such as a CD or DVD that is not included in the version you purchased, you may download this material at http://booksupport.wiley.com. For more information about Wiley products, visit www.wiley.com.

Library of Congress Cataloging-in-Publication Data:

Davis, Thomas Bieber, 1942-
    Audel industrial multi-craft mini-ref / Thomas B. Davis.—1st ed.
        p. cm.—(Audel technical trades series; 47)

    Includes index.

ISBN 978-1-118-01594-0 (pbk.), ISBN 978-1-118-13131-2 (ebk);
ISBN 978-1-118-13132-9 (ebk); ISBN 978-1-118-13133-6 (ebk);
ISBN 978-1-118-14118-2 (ebk); ISBN 978-1-118-14119-9 (ebk);
ISBN 978-1-118-14120-5 (ebk)

    1. Machinery—Handbooks, manuals, etc.   2. Mechanical engineering—
Handbooks, manuals, etc.   3. Shop mathematics—Handbooks, manuals, etc.
I. Title.   II. Title: Audel multicraft industrial reference.

    TJ151.D38 2011
    621—dc23

                                                    2011024784

Printed in the United States of America

10 9 8 7 6 5 4 3 2 1

CONTENTS

PREFACE

The Audel line of books has been a mainstay for mechanics, millwrights, machinists, electricians, and engineering personnel since the early 1900s. This author has a personal collection of many of the early Audel texts. While reading books composed in the period from 1913 to 1928, it is interesting to note that it was not the degreed mechanical or electrical engineer who calculated the shaft fit of a bearing, the belt length for a drive, or the voltage required for a circuit. A knowledge of math was expected of craftsmen. They did the calculations and also possessed the craftsmanship and skill to effect changes in machinery and circuitry to keep the process running.

In today's intensely competitive world, the old adage "Jack of all trades, master of none" is no longer a truth. Many facilities depend on a new type of employee, the *multicraftsman,* who possesses in-depth knowledge of many crafts and can apply that knowledge to keep things running. Where you work now, you may discover that the people who keep their jobs are those who understand that "survival of the fittest" means knowing more than just one trade—and being good at them.

The old Audel guidebooks stressed calculation and insisted that the craftsman be able to figure things out on paper. But the old books did something that newer books seem to have forgotten. They used *worked-out* examples so workers could relate the concepts to their jobs. It's easy to learn a formula in class or remember it for a test, but how about weeks or months later? At that time, the review of a worked-out example really helps to make things happen.

Too many small books provide tables, charts, and formulas but never show how to use them, nor do they provide concrete examples of how this information can be used on the job to make the craftsman faster, better, and more sure that the job will be a success.

Books in the Audel Mini-Ref Series are geared to the craftsman who wants to learn individually outside of a classroom as well as the tradesman who was taught but needs a few memory joggers to remember what was learned earlier. The Mini-Ref Series uses real examples to show how to make the calculation, how to use tables to find the right answer, and how to read graphs to check the theoretical against the real-life situation.

ABOUT THE AUTHOR

Thomas (Tom) Bieber Davis owns and operates Maintenance Trouble-shooting, an engineering business dedicated to diagnosis and repair of mechanical rotating equipment. For the past 40 years, Tom has worked in all 50 states in various industries. Tom's mantra, "Machinery does not understand politics, only the laws of physics," explains his insight into how rotating equipment works, as well as how it fails. As a practicing mechanical engineer, Tom enjoys teaching on-site classes of mechanical personnel about pumps, fans, and bearings, as well as writing on these subjects. Tom says his mother, Verna Bieber Davis, was the first multi-craftsman he ever knew. She could do anything. He can easily be reached by e-mail at mechanicalengineer@pobox.com, and he answers his mail.

MACHINING

DRILLING

Drilling holes in materials seems easy, but obtaining a clean, smooth hole without breaking the drill bit takes only a bit of knowledge about drilling. Here are some quick rules to make every drilling job better.

The tip of a drill is *not* a small point but a horizontal lip. A center punch must be used to make sure the drill starts at the desired position on the workpiece. If a large drill is used, it is very important to drill a pilot hole first in order to achieve proper centering.

The *smaller* the drill, the faster the speed at which it needs to turn, and the slower the feed.

The *larger* the drill, the slower the speed at which it needs to turn, and the faster the feed.

Quick Guide to Drill Press Speed (RPM)

Drill Size	Steel	Brass	Aluminum*	Plastic (Acrylic)	Hard Wood	Soft Wood	Cast Iron
1/8″–3/16″	3000	3000	3000	2500	3000	3000	1500
1/4″–3/8″	1000	1200	2500	2000	1500	3000	700
7/16″–5/8″	600	750	1500	1500	750	1500	500
11/16″–1″	350	400	1000	1000	500	750	400

* When drilling aluminum, use a "pecking" technique, bringing the drill down and back up frequently, to remove chips and keep flutes from clogging. If this is not done, drill breakage can result.

DRILL PRESS TROUBLESHOOTING

Symptoms	Cause	Fix
Drill breaks or snaps.	Speed is too low in proportion to the feed. Drill is dull. Lip clearance is too small.	Increase speed or decrease feed. Sharpen drill. Regrind properly.
Outer corners of cutting edges are breaking down.	Material being drilled has hard spots. Speed is excessive. Improper cutting oil is being used. Lubricant is not reaching point of drill.	Reduce speed. Use different lubricant and apply more often. Remove drill and spray hole.
Drill breaks when drilling wood or brass.	Chips are clogging flutes.	Increase speed. Use drills designed for these materials.
Hole is too large.	Angle or length of cutting edges (or both) is unequal, Spindle on drill press is loose.	Regrind drill properly. Tighten or adjust spindle.
Drill fails to cut stainless steel, although initially it worked okay.	Stainless steel has work-hardened because of slow feed and drill no longer cuts because cutting edges are dulled.	Resharpen drill and use heavy feed to cut through work-hardened surface, then use increased feed to continue.
Drill overheats on deep hole.	Heat cannot be removed because drill is going too fast.	Reduce speed and remove drill from hole frequently.

DRILL SHARPENING

There are a great many gadgets on the market to sharpen drills, but freehand sharpening is a skill that is easy to master and can be used to make drilling certain materials a lot easier. Sharpening a drill differently for steel or brass or aluminum can make a hard job easy and give a clean, burr-free hole.

The most common parts of a twist drill are shown here.

A drill is really a rotating chisel called the *lip* (or cutting edge), which skives off material as it turns. The *flutes* allow the material a place to go so that more material can be cut.

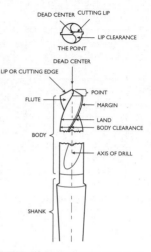

Drill Tip Geometry	
Material	**Point Angle**
Aluminum	90–135°
Brass	90–118°
Cast Iron	90–118°
Mild Steel	118–135°
Stainless Steel	118–135°
Plastics	60–90°

The drill-point angle can be changed to better accommodate drilling holes in various materials. The standard drill bit has a 118-degree included angle and will work in most materials.

There are generally three essential requirements in twist drill sharpening:

1. Equal drill-point angles, which are usually 59 degrees each, for a total of 118 degrees
2. Cutting lips of equal length
3. Correct clearance behind the cutting lips, which is approximately 8 to 12 degrees

Grind all twist drills without overheating them. Keep their points cool enough so that you can touch them with your bare fingers. Making very light passes over the wheel can do this. If a drill is under 1/8″ in diameter, the rule of thumb is to just get a new one rather than sharpening.

When sharpening, always make sure that there is relief behind the cutting edge. If the trailing edge of the bit is higher than the cutting edge, it will not cut.

If the bit is used to drill hard material, the end should be flatter so it won't dig in and dull fast. *Always* center-punch the work to prevent the drill from "walking." The 59-degree angle used on many drills is a standard angle for mild steel; for softer materials, make the point more like a pencil. When learning, start with a large bit, and remember that if the steel gets hot enough to turn black at the cutting edge, carbon has been removed from the metal and the edge will dull *very* soon. Grind slowly, and dip in water often. Get a new bit and compare the shape to the ground one—there should be very little difference. When thinning the web, remember that the thinner the web, the more easily it will penetrate the work, but also the more easily it can grab and split the drill down the center, ruining the drill immediately. If the sides (lands) are worn on the drill bit, it will grab and break.

After grinding, hold the bit up with a bright background behind it and with the cutting edges going left and right. Make sure that both the tips are the same height and that the point is in the center before attempting to thin the web (if needed). If there is a pilot hole with a greater diameter than the thickness of the web, no thinning is required.

When grinding by hand, always hold the cutting lip level against the wheel at the center height of the wheel and rotate the drill upward to make the relief. Grind slowly and don't try it with a wheel that is out of round—you will get hurt badly. Dress the wheel to be true before starting any grinding process.

Begin by holding the bit on your forefinger, with its cutting lip horizontal and the axis of the drill at an angle of about 59 degrees. The actual grinding process involves three distinct motions of the shank while the bit is held lightly against the wheel. The three motions occur simultaneously:

- ■ To the left
- ■ In a clockwise rotation
- ■ Downward

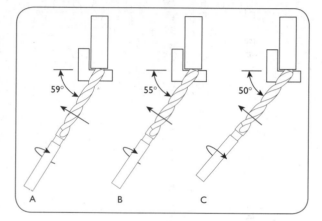

DRILL SIZES AND DECIMAL EQUIVALENTS

Drill	mm	Decimal	Drill	mm	Decimal
–	0.10	.0039	62	0.97	.0370
–	0.20	.0079	61	0.99	.0380
–	0.25	.0098	–	1.00	.0390
–	0.30	.0118	60	1.02	.0400
80	0.34	.0135	59	1.04	.0410
79	0.37	.0145	58	1.07	.0420
1/64	0.40	.0156	57	1.09	.0430
78	0.41	.0160	56	1.18	.0465
77	0.46	.0180	3/64	1.19	.0469
–	0.50	.0197	55	1.32	.0520
76	0.51	.0200	54	1.40	.0550
75	0.53	.0210	53	1.51	.0595
74	0.57	.0225	1/16	1.59	.0625
–	0.60	.0236	52	1.61	.0635
73	0.61	.0240	51	1.70	.0670
72	0.64	.0250	50	1.78	.0700
71	0.66	.0260	49	1.85	.0730
–	0.70	.0276	48	1.93	.0760
70	0.71	.0280	5/64	1.98	.0781
69	0.74	.0292	47	1.99	.0785
–	0.75	.0295	–	2.0	.0787
68	0.79	.0310	46	2.06	.0810
1/32	0.79	.0313	45	2.08	.0820
–	0.80	.0315	44	2.18	.0860
67	0.81	.0320	43	2.26	.0890
66	0.84	.0330	42	2.37	.0935
65	0.89	.0350	3/32	2.38	.0938
–	0.90	.0350	41	2.44	.0960
64	0.91	.0354	40	2,50	.0980

Drill	mm	Decimal	Drill	mm	Decimal
63	0.94	.0360	39	2.53	.0995
38	2.50	.1015	15	4.57	,1800
37	2.64	.1040	14	4.62	.1820
36	2.71	.1065	13	4.70	.1850
7/64	2.78	.1094	3/16	4.76	.1875
35	2.79	.1100	12	4.80	.1890
34	2.82	.1100	11	4.85	.1910
33	2.87	.1130	10	4.91	.1935
32	2.95	.1160	9	4.98	.1960
–	3.00	.1181	–	5.00	.1968
31	3.05	.1200	8	5.05	.1990
1/8	3.18	.1250	7	5.11	.2010
30	3.26	.1285	13/64	5.16	.2031
29	3.45	.1360	6	5.18	.2040
28	3.57	.1405	5	5.22	.2055
9/64	3.57	.1406	4	5.31	.2090
27	3.66	.1440	3	5.41	.2130
26	3.73	.1470	7/32	5.56	.2188
25	3.80	.1495	2	5.61	.2210
24	3.86	.1520	1	5.79	.2280
23	3.91	.1540	A	5.94	.2340
5/32	3.97	.1562	15/64	5.95	.2344
22	3.98	.1570	–	6.00	.2362
–	4.00	.1575	B	6.05	.2380
21	4.04	.1590	C	6.15	.2420
20	4.09	.1610	D	6.25	.2460
19	4.22	.1660	1/4	6.35	.2500
18	4.31	.1695	E	6.35	.2500
11/64	4.37	.1719	F	6.53	.2570
17	4.39	.1730	G	6.63	.2610
16	4.50	.1770	17/64	6.75	.2656

Drill	mm	Decimal	Drill	mm	Decimal
H	6.76	.2660	Z	10.49	.4130
I	6.91	.2720	27/64	10.72	.4219
–	7.00	.2756	–	11.00	.4331
J	7.04	.2770	7/16	11.11	.4375
K	7.14	.2810	29/64	11.51	.4531
9/32	7.14	.2812	15/32	11.91	.4688
L	7.37	.2900	–	12.00	.4724
M	7.49	.2950	31/64	12.30	.4844
19/64	7.54	.2969	1/2	12.70	.5000
N	7.67	.3020	–	13.00	.5118
5/16	7.94	.3125	33/64	13.10	.5156
–	8.00	.3150	17/32	13.49	.5312
O	8.03	.3160	35/64	13.89	.5469
P	8.20	.3230	–	14.00	.5512
21/64	8.33	.3281	9/16	14.29	.5625
Q	8.43	.3320	37/64	14.68	.5781
R	8.61	.3390	–	15.00	.5906
11/32	8.73	.3438	19/32	15.08	.5938
S	8.84	.3480	39/64	15.48	.6094
–	9.00	.3543	5/8	15.88	.6250
T	9.09	.3580	–	16.00	.6299
23/64	9.13	.3594	41/64	16.27	.6406
U	9.35	.3680	21/32	16.67	.6562
3/8	9.53	.3750	–	17.00	.6693
V	9.56	.3770	43/64	17.07	.6719
W	9.80	.3860	11/16	17.46	.6875
25/64	9.92	.3906	45/64	17.86	.7031
–	10.00	.3937	–	18.00	.7087
X	10.08	.3970	23/32	18.26	.7188
Y	10.26	.4040	47/64	18.65	.7344
–	19.00	.7480	7/8	22.23	.8750
3/4	19.05	.7500	57/64	22.62	.8906

Drill	mm	Decimal	Drill	mm	Decimal
49/64	19.45	.7656	–	23.00	.9055
25/32	19.84	.7812	29/32	23.02	.9062
–	20.00	.7874	59/64	23.42	.9219
51/64	20.24	.7969	15/16	23.81	.9375
13/16	20.64	.8125	–	24.00	.9449
–	21.00	.8268	61/64	24.21	.9531
53/64	21.03	.8281	31/32	24.61	.9688
27/32	21.43	.8438	–	25.00	.9842
55/64	21.84	.8594	63/64	25.00	.9844
–	22.00	.8661	1″	25.40	1.0000

Type of Drill	Characteristics
Low-carbon steel	Cheap. Used for wood. Requires frequent sharpening.
High-carbon steel	Can be used on metal or wood. Loses temper if overheated.
High-speed steel	A form of tool steel. Resistant to heat. Good for wood and various metals. General purpose.
Cobalt steel	Holds hardness at higher temperatures. Works with stainless steel.
Black oxide and titanium nitride	Coatings used on drills to improve hardness. Unfortunately, removed by sharpening.

SCREW THREADS

Recorded history shows the screw thread was developed by Archimedes in the third century BC. He used a pipe wrapped around a shaft in a helical pattern to make a crude bilge pump for ships. Archimedes took the basic inclined plane and wrapped it into a spiral shape. Early screws were made by wrapping wire around plain bar. Nuts were made of softer material (copper, for example) by forging them around the wire-wrapped rod. Later, screws were cut from solid bar using single-point cutting tools or chasers. Because of its many advantages, thread rolling is the preferred method of manufacture today. Early screw manufacturing suffered from the absence of accurate and powerful machinery capable of holding minimally accurate tolerances. This was compounded by the lack of accurate inspection methods. For many years screws and nuts were manufactured and used in matched sets and, as a result, were not interchangeable.

Most mechanical screw thread has three different pitches (threads per inch) available for a given diameter. The terms used to refer to them are *coarse, fine,* and *extra-fine.*

Coarse thread (UNC) is used for bolts, screws, nuts, and other applications where rapid assembly or disassembly is needed or corrosion or other slight damage might occur.

Fine thread (UNF) is used for bolts, screws, nuts and other applications where a finer thread is needed. It is used where the length of engagement is short, where a smaller lead angle is desired, or when the wall thickness demands a fine pitch.

Extra-fine thread is used where even finer pitches are desirable for a short length of engagement and for thin-walled tubes, nuts, ferrules, or couplings.

Machine screws with a body diameter less than 1/4″ use a number designation ranging from #0 to #12. The major diameter of #0 is 0.0600″ and each higher number adds 0.0130″ to the previous diameter. The major diameter of #3 screw thread is 0.0600″ + 0.0130″ + 0.0130″ + 0.0130″, or 0.0990″.

Fractional screw sizes are generally limited to ¼ through 5/8 (by 1/16ths), ¾ through 1-¼ (by 1/8ths), and 1-½ through 2 (by ¼ths). They are commonly available in UNC and UNF pitches. UNEF pitches are available, but they are not commonly stocked.

You don't have a thread chart available and you need to know the major diameter of a #8 machine screw? Remember that the #0 screw is 0.0600″, and add 0.0130″ for each number designation. Starting with the #0 screw and a major diameter of 0.0600″, add 0.0130″ × 8 = 0.1040″. 0.0600″ + 0.1040″ = 0.1640″. How about a #6 screw? Answer: 0.1380″

Thread standards for pipe include National Pipe Tapered (NPT) and National Pipe Straight (NPS). The threads for tapered pipe are cut on a taper of 1/16″ per inch. This taper allows them to form a seal when torqued as the flanks of the threads compress against each other, as opposed to NPS fittings in which the threads merely hold the pieces together and do not provide the seal. However, a clearance remains between the crests and the roots of the threads, resulting in leakage around this spiral. This means that NPT fittings must be made leakfree with the aid of thread seal tape or a thread sealant.

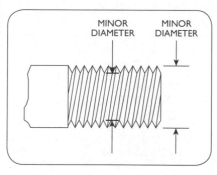

DRILLING AND TAPPING A HOLE

Drilling and tapping a hole with threads can be a tricky operation. Good taps are made of hardened high-speed steel (HSS) to allow sharp cutting of the threads, but they also are susceptible to breaking or snapping if care is not used. The hole drilled for a tapped thread is usually large enough to allow about 75 percent threading. This amount is adequate to keep the threads from stripping out while tightening a bolt or fastener to a recommended torque. One of the most common reasons taps break is that the drilled hole is too small and the threading requirement is above 75 percent. This causes excessive torque on the tap handle. The tap is very likely to snap. Since a tap is harder than most drills, a broken tap is difficult to drill out using conventional drills. Another reason for tap breakage is that, of all the machining operations, tapping requires the most lubricant. Too little lubricant causes the tap to bind and snap. You cannot use enough lubricant when you are tapping.

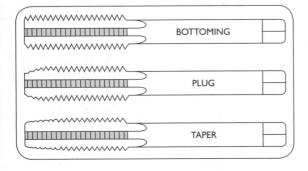

There are three types of taps in common usage: the plug tap, the bottoming tap, and the taper tap. The plug tap is used for production tapping with a tapping machine. Most maintenance work involves the use of the bottoming and taper types of taps. The taper tap is easy to start and keep straight to make sure the threads are at right angles to the axis of the drilled hole. The bottoming tap is used when a hole is drilled into a piece of material but not all the way through. Usually a hole must be drilled a little farther than the length of threads needed with a bottoming tap. If a plug or taper tap is used in a blind hole, there is a good chance it will lock up or break.

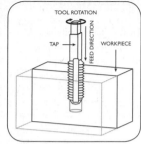

Bottoming taps should not be used to start a new thread. Use a taper tap to begin the threading, and then remove and use the bottoming tap to finish a blind hole.

Taps are usually broken by bending, not twisting. A wrench that pushes on only one side puts an asymmetrical stress on the tap and is very likely to break it. Instead, a "T" handle should always be used, which makes it easier to turn the tap without bending it.

Sometimes when tapping a hole, commercially produced lubricants aren't available. In that case, the following lubricants may be substituted:

Material	Substitute Lubricant
Carbon steel	Sulfurized cutting oil used for pipe threading and found in hardware stores
Aluminum	Kerosene or fuel oil
Cast iron	Can be dry-tapped or add some WD-40
Plastics	Dry-tap or use liquid soap
Copper	Crisco shortening

The steps involved in tapping a hole are as follows:

1. Determine the size and depth of the thread desired; usually, this is the same size as the bolt or fastener being screwed into the hole.
2. Determine the tap drill size from a chart or calculation.
3. Drill the hole to the correct depth. Remember that taps cannot thread all the way to the bottom of a blind hole, so a blind hole must be deeper than the thread depth.

4. Blow out the hole to remove any drill shavings.
5. Make sure the workpiece is locked down or braced to prevent it from moving.
6. Apply enough lubricant to the hole and the tap.
7. Insert the tap into a "T" handle or similar holder.
8. Hold the tap in a straight line with the hole and turn it clockwise. The sharp cutters on the tap will bite into the material and begin forming the threads.
9. Every few turns, back the tap up (turn it counterclockwise) to clear away chips of loose material. With some materials, the tap may need to be backed up even after making a half turn. If the pressure increases on the handle, back up to clear the tool and start again. Add more lubricant.
10. When the tap reaches the proper depth or creates sufficient threads, back it all the way out, blow the chips out, and make sure to brush any chips out of the tap itself before putting it away.

Small taps are extremely easy to break. It takes only about 5 lb of force (off center) to snap a #6-32 tap. Most better sets of taps are necked down at the top to confine breakage to this area, allowing the tap to be removed with a vise-grip pliers, since part of it protrudes above the workpiece. If the workpiece is extremely valuable, a broken tap can be removed with electrical discharge machining (EDM) techniques, usually available at any precision machine shop. The price for removing the tap is much less than it would cost to scrap an expensive part.

CALCULATING A TAP DRILL SIZE WITHOUT A TABLE

If you are without a table for selecting the proper tap drill size, it can be calculated quickly.

Subtract 1/pitch (threads per inch) from the major diameter of the thread. A few examples show how to do this:

Drill and tap for 1/4-20 NC threads

1/4″ = major diameter of the thread

20 = pitch, or threads per inch

1/4″ (0.250″) − 1/20″ (0.050″) = 0.200″

Either a #7 (0.201″) or a 13/64″ (0.203″) drill will work.

Drill and tap for 1/4-28 NF threads

1/4″ = major diameter of the thread

28 = pitch, or threads per inch

1/4″ (0.250″) − 1/24″ (0.035″) = 0.215″

Either a #3 (0.213″) drill or a 7/32″ (0.218″) drill will work.

You must drill and tap a 1/2-13 NC threaded hole. What tap drill size do you need? 1/2″ = 0.500″. 1/13 = 0.076″. 0.500″ − 0.076″ = 0.424″. 27/64″ (0.421″) is the tap drill size needed.

Calculate the tap drill size for a 3/8-16 NC drilled and tapped hole.

pəsn əq pɪnoɥs ɪɪɪɹp dɐʇ ,,9ɪ/ϛ ɐ ɹo ,,ZɪƐ˙0 = ,,Ɛ90˙0 − ,,ϛϟƐ˙0

,,Ɛ90˙0 = ,,9ɪ/ɪ puɐ ,,ϛϟƐ˙0 = ,,8/Ɛ

What about a 5/8-11 NC drilled and tapped hole?

·pəsn əq pɪnoɥs ɪɪɪɹp dɐʇ ,,ZƐ/ϟɪ ɐ ɹo ,,ϛƐϛ˙0 = ,,060˙0 − ,,ϛZ9˙0

,,060˙0 = ,,ɪɪ/ɪ puɐ ,,ϛZ9˙0 = ,,8/ϛ

SCREW THREAD—DRILL AND TAP CHART (UNC AND UNF)

Machine Screw Size		Threads per Inch	Minor Diameter	Tap Drills				Clearance Hole Drills			
				75% Thread		50% Thread		Close Fit		Free Fit	
# or Diam.	Major Diam.			Drill Size	Decimal Equiv.	Drill Size	Decimal Equiv.	Drill Size	Decimal Equiv.	Drill Size	Decimal Equiv.
0	.0600	80	.0447	3/64	.0469	55	.0520	52	.0635	50	.0700
1	.0730	64	.0538	53	.0595	1/16	.0625	48	.0760	46	.0810
		72	.0560	53	.0595	52	.0635				
2	.0860	56	.0641	50	.0700	49	.0730	43	.0890	41	.0960
		64	.0668	50	.0700	48	.0760				
3	.0990	48	.0734	45	.0785	44	.0860	37	.1040	35	.1100
		56	.0771	43	.0820	43	.0890				
4	.1120	40	.0813	43	.0890	41	.0960	32	.1160	30	.1285
		48	.0864	42	.0935	40	.0980				
5	.1250	40	.0943	38	.1015	7/64	.1094	30	.1285	29	.1360
		44	.0971	37	.1040	35	.1100				
6	.1380	32	.0997	36	.1065	32	.1160	27	.1440	25	.1495
		40	.1073	33	.1130	31	.1200				
8	.1640	32	.1257	29	.1360	27	.1440	18	.1695	16	.1770
		36	.1299	29	.1360	26	.1470				
10	.1900	24	.1389	25	.1495	20	.1610	9	.1960	7	.2010
		32	.1517	21	.1590	18	.1695				
12	.2160	24	.1649	16	.1770	12	.1890	2	.2210	1	.2280
		28	.1722	14	.1820	10	.1935				
1/4	.2500	20	.1887	7	.2010	7/32	.2188	F	.2570	H	.2660
		28	.2062	3	.2130	1	.2280				
5/16	.3125	18	.2443	F	.2570	J	.2770	P	.3230	Q	.3320
		24	.2614	I	.2720	9/32	.2812				
3/8	.3750	16	.2983	5/16	.3125	Q	.3320	W	.3860	X	.3970
		24	.3239	Q	.3320	S	.3480				
7/16	.4375	14	.3499	U	.3680	25/64	.3906	29/6	.4531	15/32	.4687
		20	.3762	25/64	.3906	13/32	.4062				
1/2	.5000	13	.4056	27/64	.4219	29/64	.4531	33/6	.5156	17/32	.5312
		20	.4387	29/64	.4531	15/32	.4688				
9/16	.5625	12	.4603	31/64	.4844	33/64	.5156	37/6	.5781	19/32	.5938
		18	.4943	33/64	.5156	17/32	.5312				
5/8	.6250	11	.5135	17/32	.5312	9/16	.5625	41/64	.6406	21/32	.6562
		18	.5568	37/64	.5781	16/32	.5938				
3/4	.7500	10	.6273	21/32	.6562	11/16	.6875	49/6	.7656	25/32	.7812
		16	.6733	11/16	.6875	45/64	.7031				
7/8	.8750	9	.7387	49/64	.7656	51/64	.7969	57/6	.8906	29/32	.9062
		14	.7874	13/16	.8125	53/64	.8281				
1	1.000	8	.8466	7/8	.8750	59/64	.9219	1-	1.0156	1-1/32	1.0313
		12	.8978	15/16	.9375	61/64	.9531				

TAPER PIPE—DRILL AND TAP CHART (NPT)

Tap Size	Threads per Inch	Tap Drill
1/8	27	11/32
1/4	18	7/16
3/8	18	37/64
1/2	14	23/32
3/4	14	59/64
1	11-1/2	1-5/32
1-1/4	11-1/2	1-1/2
1-1/2	11-1/2	1-3/4
2	11-1/2	2-7/32
2-1/2	8	2-21/32
3	8	3-1/4
3-1/2	8	3-3/4
4	8	4-1/4

SCREW THREAD—DRILL AND TAP CHART (UNEF)

Machine Screw Size				Tap Drills				Clearance Hole Drills			
				75% Thread		50% Thread		Close Fit		Free Fit	
# or Diameter	Major Diameter			Drill Size	Decimal Equivalent	Drill Size	Decimal Equivalent	Drill Size	Decimal Equivalent	Drill Size	Decimal Equivalent
12	.2160	32	.1777	13	.1850	9	.1960	2	.2210	1	.2280
1/4	.2500	32	.2117	7/32	.2188	1	.2280	F	.2570	H	.2660
5/16	.3125	32	.2742	9/32	.2812	L	.2900	P	.3230	Q	.3320
3/8	.3750	32	.3367	11/32	.3438	T	.3580	W	.3860	X	.3970
7/16	.4375	28	.3937	Y	.4040	Z	.4130	29/64	.4531	15/32	.4687
1/2	.5000	28	.4562	15/32	.4688	31/64	.4844	33/64	.5156	17/32	.5312
9/16	.5625	24	.5114	33/64	.5156	17/32	.5312	37/64	.5781	19/32	.5938
5/8	.6250	24	.5739	37/64	.5781	19/32	.5938	41/64	.6406	21/32	.6562
11/16	.6825	24	.6364	41/64	.6406	21/32	.6562	45/64	.7031	23/32	.6562
3/4	.7500	20	.6887	45/64	.7031	23/32	.7188	49/64	.7656	25/32	.7812
13/16	.8125	20	.7512	49/64	.7656	25/32	.7812	53/64	.8281	27/32	.8438
7/8	.8750	20	.8137	53/64	.8281	27/32	.8438	57/64	.8906	29/32	.9062
15/16	.9375	20	.8762	57/64	.8906	29/32	.9062	61/64	.9531	31/32	.9688
1	1.000	20	.9387	61/64	.9531	31/32	.9688	1-1/64	1.0156	1-1/32	1.0313

METALWORKING LUBRICANTS

Material	Type of Operation		
	Drilling	**Threading**	**Milling or Turning**
Aluminum	Soluble oil Kerosene	Soluble oil Kerosene	Soluble oil
Brass	Dry Soluble oil Kerosene	Soluble oil	Soluble oil
Bronze	Dry Soluble oil	Soluble oil	Soluble oil
Cast Iron	Dry Air jet Soluble oil	Dry Sulfurized oil	Dry Soluble oil
Copper	Dry Soluble oil Kerosene	Lard oil	Soluble oil
Machine Steel	Soluble oil Sulfurized oil Mineral lard oil	Soluble oil Mineral lard oil	Soluble oil
Malleable Iron	Dry Soda water	Lard oil Soda water	Soluble oil Soda water
Monel	Soluble oil	Soluble oil	Lard oil
Steel Alloys	Soluble oil Sulfurized oil	Sulfurized oil	Soluble oil
Tool Steel	Soluble oil Sulfurized oil	Sulfurized oil	Soluble oil

2

METALS

THERMAL EXPANSION OF METALS

It is a common practice to use shrink fits (often called interference fits) for affixing one machinery part to another. The knowledge and ability to calculate a shrink fit for a coupling onto a shaft or reduction of the diameter of a shaft with low temperatures to allow a similar attachment are based on a simple formula that involves multiplication, temperature difference, measurement of the original diameter, and the linear coefficient of expansion of the materials involved.

The formula is often called the "pipe fitter's growth formula" in the trades. Pipe fitters use it to figure out the increase in length of pipes due to hot fluids passing through. The calculation of growth tells the fitter when and where to install expansion joints so the piping does not end up looking like the back of a sea serpent when it's heated up. Imagine the growth that must be considered for the 800-mile-long Alaska pipeline!

$$\Delta d = d \times \Delta t \times l_e$$

Δd = amount of growth (or shrinkage)
d = original dimension (length of pipe, diameter, etc.)
Δd = change in temperature for the two conditions
l_e = linear coefficient of expansion of the material used

Imagine a coupling with a 1.998″-diameter hole that must be shrunk onto the end of a shaft that measures 2.000″. Obviously, the shaft is bigger than the hole, and an interference or shrink fit will be involved. The pipe

19

fitter's formula can be used to determine the growth of the hole if the coupling is heated to a temperature of 350°F using a torch.

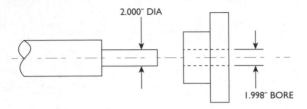

For this example, imagine that the shaft and the coupling are made of steel. The linear coefficient of expansion of steel is 0.0000063 in/in/°F. This means that a 1″ length of steel will grow 0.0000063″ if it is heated 1°F.

$$\Delta d = d \times \Delta t \times l_e$$

Δd = amount of growth (or shrinkage)

d = original dimension of the coupling = 1.998″ bore

Δd = assuming the area around the machine is 70°F and the coupling is heated to 350°F, the difference in the temperatures is (350°F − 70°F) = 280°F

l_e = linear coefficient of steel = 0.0000063 in/in/°F

Substituting into the formula yields:

$$\Delta d = 1.998 \times 280 \times 0.0000063$$
$$\Delta d = 0.0035″$$

An increase of 0.002″ (2.000″ − 1.998″) was needed to "just" tap the coupling onto the shaft. That might be a little difficult because if the shaft heats up a tiny bit or the coupling cools down the job gets tough. Heating to 350°F gives an increase in the bore of 0.0035″, making the bore 2.0015″—almost 0.002′ of clearance. An easy assembly for sure!

LINEAR EXPANSION FOR VARIOUS MATERIALS

Linear Expansion for Various Materials	
Material	l_e = **linear coefficient of expansion, in/in/°F**
ABS plastic	0.0000410 in/in/°F
Acrylic (extruded)	0.0001300 in/in/°F
Acrylic (cast sheet)	0.0000410 in/in/°F
Aluminum	0.0000123 in/in/°F
Brass	0.0000096 in/in/°F
Bronze	0.0000098 in/in/°F
Copper	0.0000089 in/in/°F
Glass	0.0000050 in/in/°F
Gold	0.0000079 in/in/°F
Iron, cast	0.0000056 in/in/°F
Iron, wrought	0.0000065 in/in/°F
Lead	0.0000157 in/in/°F
Nickel	0.0000695 in/in/°F
Nylon	0.0000045 in/in/°F
Polycarbonate	0.0000440 in/in/°F
Polyethylene	0.0001110 in/in/°F
PVC	0.0000730 in/in/°F
Silver	0.0000110 in/in/°F
Steel	0.0000063 in/in/°F
Tin	0.0000116 in/in/°F
Titanium	0.0000047 in/in/°F
Zinc	0.0000141 in/in/°F

The thermal expansion formula can be expressed another other way:

$$\Delta t = \frac{\Delta d}{d \times l_e}$$

If the coupling in the previous example has a 1.998″ bored hole but is made of aluminum, *what temperature allows an expansion that is 0.002″ greater than the* steel *shaft diameter?*

The shaft diameter is 2.000″, and if the hole in the aluminum coupling must be expanded to 0.002″ greater than the shaft, then the hole must end up being 2.002″ after heating. That amounts to a 0.0004″ growth in the bore, since 2.002″ – 1.998″ = 0.004″. Using the table, we find the linear coefficient of expansion of aluminum is $l_e = 0.0000123$ in/in/°F.

$$\Delta t = \frac{\Delta d}{d \times l_e}$$

Δd = *amount of growth* = 0.004″
d = *original dimension of the coupling bore* = 1.998″
l_e = *linear coefficient of aluminum* = 0.0000123 in/in/°F

$$\Delta t = \frac{0.004}{1.998 \times 0.0000123}$$

$$\Delta t = 163°F$$

The temperature to which the aluminum coupling must be heated to yield an expanded bore of 2.004″ is 163°F above the ambient temperature—say, 70°F. 163°F + 70°F = 233°F. In the previous example, with a steel coupling, the temperature had to be expanded 280°F above ambient for the same expansion, since the coefficient of expansion was smaller. Less heat is required to expand the aluminum. The aluminum expands about twice as much for a given temperature as the steel.
If the coupling was made of brass, can you figure the temperature required?

Answer: Δt = 208°F gives 0.004″ expansion.

Many pumping processes involve moving a hot fluid inside a pipe. Suppose that a heated oil with a temperature of 800°F is pumped through a 100-foot-long pipe located between the pump and the discharge from the bottom of the tank. The pipe fitter makes a mistake and does not include an expansion joint in the piping. Obviously, the pipe is going to grow in length due to the increased temperature. The original piping was done in ambient temperatures, but now the pipe is hot. Using the pipe-fitting formula, this growth can be calculated. 100′ of pipe × 12″ in a foot = 1200″ of pipe. Assume an ambient temperature of 70°F. The change in temperature seen by the pipe is therefore 800°F – 70°F = 730°F. The pipe material is steel, so the l_e = 0.0000063 in/in/°F. Substituting the values in the pipe fitter's formula shows the following:

$$\Delta d = d \text{ x } \Delta t \text{ x } l_e$$
$$\Delta d = 1200 \times 730 \times 0.0000063$$
$$\Delta d = a \text{ whopping } 5.5''$$

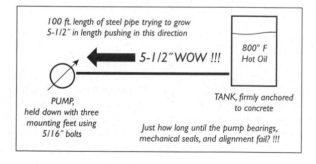

100 ft. length of steel pipe trying to grow 5-1/2″ in length pushing in this direction

5-1/2″ WOW !!!

800° F
Hot Oil

PUMP,
held down with three
mounting feet using
5/16″ bolts

TANK, firmly anchored
to concrete

Just how long until the pump bearings,
mechanical seals, and alignment fail? !!!

3

REPAIRS AND REBUILDING

STUB SHAFT REPAIR

A repair can be made to the damaged end of a shaft by removing the bad section and replacing it with a new "stub" end. The procedure involves machining, welding, and calculation of thermal expansion for fitting the stub to the original shaft. While not all facilities are equipped with the machinery necessary to perform such a repair, the knowledge of this technique may be used successfully to work with an outside maintenance/ machine shop to perform a reliable correction to a damaged shaft.

The following steps illustrate the procedure needed to stub a shaft:
1. If a blueprint is not available, make a drawing of the shaft showing all dimensions.

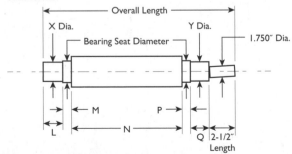

Note that this shaft is bent and damaged where the coupling would attach. The rest of the shaft is considered to be dimensionally correct and can be used again.

The drawing does not need to be beautiful. Notice how this drawing is made with simple rectangles and includes a place for both lengths and diameters critical to the remaking of the shaft.

2. Mount the undamaged end of the shaft in a four-jaw chuck and "zero in" the shaft near the jaws of the chuck. Use soft jaws or aluminum shims to prevent damage to the shaft surface.

3. Position a center rest around the shaft so the center rest is between the chuck and the damaged end of the shaft. Adjust the center rest using a dial indicator to "zero in" the shaft at the center rest position.

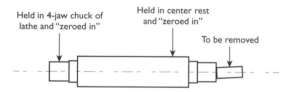

4. Cut off the damaged portion of the shaft.

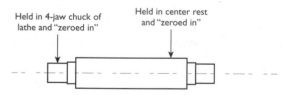

5. Face, center drill, and drill the end of the shaft. The diameter of the hole should be slightly larger than the finished stub shaft and allow for a press fit.

6. Drill a through-hole for plug welding in a noncritical part of the shaft at a right angle to the hole. Mechanical pinning may also be used if welding is not desired.

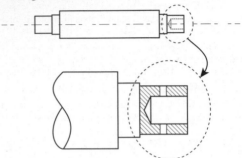

7. Make a stub of the same material as the shaft. The stub should be larger in diameter and longer than the damaged portion of the shaft plus the depth of the hole drilled in the shaft. This provides ample machining allowance.

8. Machine one end of the stub to a press-fit diameter of the hole in the shaft. The length of this portion should be slightly less than the depth of the hole in the shaft.

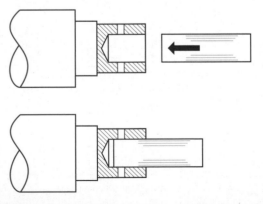

9. Welding or pinning can be used to affix the stub shaft and prevent turning or spinning.

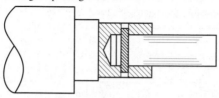

10. The shaft can be remounted and recentered in the lathe using a steady rest and a four-jawed chuck.

11. Machine the stub to the original shaft dimensions provided by the drawing or blueprint.

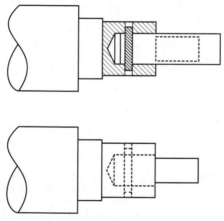

12. The shaft is repaired and can be reused.

Please note: Variations of this procedure certainly make sense. Any good maintenance machine shop or maintenance-trained machinist will understand the concept of stub shafting and can produce an effective repair, although not necessarily as depicted by these drawings. What counts is that the repair takes less time to make than making a new shaft. In a case in which downtime is critical, minutes count.

FREEING FROZEN BOLTS

Rust is the common name for iron oxide, the chemical Fe_2O_3, which is frequently found because iron and steel combine readily with oxygen. On the surface of a piece of steel, this oxide "bubbles up" to form large flakes of rust. In the process of rusting, the iron oxide expands in volume. When rust forms in the threaded area between a nut and a bolt, the rust expansion locks the threads in place. Mechanics use the term *frozen* to indicate the existence of this condition. Understanding how rust works can be helpful in determining a solution to the frozen bolt situation.

To begin, it is important to realize that any method of releasing a frozen bolt (or nut) involves breaking the bond formed within the threads by the expansion of the rust. Most craftsmen know to use a socket or box wrench to make sure that there is good contact between the wrench and the fastener. Open-end or adjustable wrenches are not going to work well in this situation. While the task at hand is to *loosen* the assembly, the real job is to break the bond holding the two parts together. Many times, *tightening the bolt or nut slightly* will break the bond. It's worth a try and should be repeated no matter what other method might be used. Beyond that tip, the methods that can be used are as follows.

Hot wax technique: Use a small propane torch to heat the bolt until it is hot enough to melt candle wax. Put a candle on top of the bolt so the bolt liquefies some of the wax. The heat will pull the liquid wax down onto the threads of the bolt. Wait until the bolt cools off a bit, then use a wrench to loosen it. Try tightening a tiny bit first.

Rust-penetrating liquid: In places where heat is not a good idea, chemicals can be used to penetrate the rust. There are various products in the marketplace, but some of the most commonly used compounds are WD-40, hydrogen peroxide, penetrating oil, and Diet Coke. Yes, Diet Coke. Any cola product will work, but Diet Coke contains the most phosphoric acid, which is the active rust-eating ingredient. It breaks the bonds of the rust and allows the threads to turn. Tapping on the bolt or nut will often break the rust bonds and allow the penetrating liquid to work into the threads. A very effective penetrating oil can be "brewed" using acetone (available in any paint store) and ATC (automatic transmission fluid) mixed in a 50-50 ratio—remarkable performance!

All penetrating liquids take time to work. One technique that works well is to use window putty or child's modeling clay to build a small cup around the bolt. Press the putty down firmly and use plenty, especially at the base, so the cup will not leak when liquid is added. Fill the cup with the penetrating solution, then go away for a while and let the liquid work. After a few hours, check to see whether the bolt can be moved. If the fastener still will not budge, use a rag to mop up whatever liquid is left in the bowl, give the bolt a whack with a hammer to jolt the metal a bit, then refill the putty cup with more of the liquid and give it a second try.

<u>Heating and cooling:</u> Using an oxy-fuel torch and a small tip, heat the bolt or nut until it just starts to redden. Then use an ice pouch or ice wrapped in a towel to rapidly cool the fastener. The heating and cooling will break the rust bond and allow penetrating oil to be used more effectively.

REMOVING BROKEN BOLTS

Just when you thought the job was going to be easy, use a little too much torque, and the bolt breaks. Now what? Here are some shop techniques that work well, arranged from the easiest problem to more difficult breaks.

Vise-grip pliers: This method works on a clean, unseized broken bolt with a reasonable amount of shank/body protruding. Lock the pliers tightly and slowly work the bolt out.

Narrow center punch: This works for situations where the bolt is broken off nearly flush or even partly below the surface. A small, sharp center punch, gently tapped with a small hammer, is very effective for winding out a broken bolt far enough to get a grip on it. This technique takes a bit of practice but, once learned, it is very useful.

Easy-out extractor: If a bolt is broken off too far below the surface to use a punch or if it is quite solidly seized, then an extracting tool, such as an easy-out, may be the solution. An easy-out is a tapered tool with four ridges running in a gentle left-hand spiral down the length. Easy-outs are normally purchased in a pack of four or five different sizes, to suit different sizes of bolts. Each tool is marked with a drill size. A hole of that drill size is drilled down the center of the broken bolt. The easy-out is *gently* tapped into the hole so that it bites into the sides. Then, using a tap wrench, the easy-out is turned counterclockwise, and the bolt winds out. Never use an adjustable wrench or open-end wrench on the square drive end of the easy-out because the easy-out can snap. They are made of very hard material and, if they break, they cannot be drilled out. Use of the tap wrench keeps forces balanced and reduces the chances of snapping.

Accurate drilling: It is possible to drill out the center of the bolt, using progressively larger drill bits, until what is left looks like a spring in the hole, which can be pried out with relative ease. In fact, quite often the heat of drilling frees the bolt. Begin with a center punch and hammer and locate the punch mark as close to the center of the bolt as possible. Be very accurate. If the drilling starts wandering away from the center when you are using this method, *stop* immediately!

Continuing will damage the parent thread. This may be the time to try an easy-out.

Drill out and retap oversize: This method involves drilling the bolt out, thread and all. It may be possible to then tap the hole to the next size up. An alternative is to fit a HeliCoil™, a device that looks a bit like a spring and is made to be threaded into an oversize hole but retain the original thread size on the inside diameter of the coil.

Weld a nut on: Place a nut over the broken bolt and, using an oxy-acetylene torch and a brazing rod, heat the bolt through the hole in the nut and add the brazing rod to weld the bolt to the nut. Slowly build up material in the center of the hole. Allow the assembly to cool and, with a socket or box wrench, remove the bolt using the new "head"—the nut.

Spark erosion (EDM): If a machine component is valuable and must be salvaged, a spark eroder (electrical discharge machining) can be used to "burn out" a bolt, broken tap, or snapped easy-out. Machine shops specializing in tool-and-die work have this equipment available. A few hundred dollars spent for EDM might be well worth it if the components are hard to find, have a long delivery time, or are expensive.

4

MACHINERY INSPECTION AND MEASUREMENT

READING A MICROMETER

Many jobs involve the use of precision measurement. Measuring a part to determine wear, checking a shaft journal to make sure it properly fits a bearing, and checking the length of a part all involve the use of a micrometer.

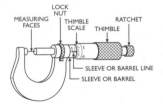

The object of measurement is placed between the measuring faces, and the ratchet or thimble is turned until the measuring faces touch the part. If using the thimble, you must have a "feel" for the contact—neither too tight nor too loose. A micrometer equipped with a ratchet eliminates the need to feel the contact. When the two faces contact the part, the ratchet slips (clicks).

The locknut may be turned when the reading is obtained to allow the micrometer to be passed around to others or carried to an area with better lighting in order to read the values shown.

Some micrometers have an additional scale on the top of the sleeve called a *vernier* scale. This scale allows these micrometers to read to the 0.0001 (ten–thousandth) of an inch required for assembly of antifriction bearings.

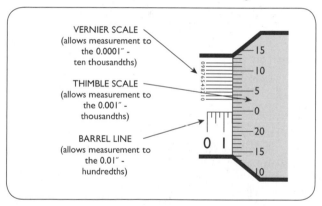

VERNIER SCALE
(allows measurement to
the 0.0001″ -
ten thousandths)

THIMBLE SCALE
(allows measurement to
the 0.001″ -
thousandths)

BARREL LINE
(allows measurement to
the 0.01″ -
hundredths)

The barrel line is divided into hundredths, and each large number indicates 0.100″ of space. The micrometer shown here has opened at least 0.100″. The lines between each large number on the barrel are in increments of 0.025″ ($4 \times 0.025″ = 0.100″$—makes sense, right?). The thimble scale divides each small barrel line increment into 25 parts, so each increment on the thimble is 0.001″. Let's ignore the vernier scale for a minute.

Looking at the same drawing, the following addition of items gives the reading:

0.100″	—Large number barrel reading
+ 0.025″	—One increment on barrel
+ 0.025″	—Another increment on barrel
+ 0.000″	—Increment on thimble
0.150″	—Total reading

Let's take a look at a new micrometer reading and again write the answer down.

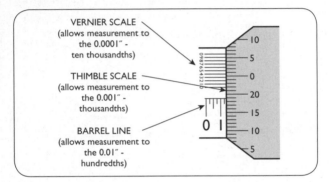

The barrel line is divided into hundredths, and each large number indicates 0.100″ of space. The micrometer shown here has opened at least 0.100″. The lines between each large number on the barrel are in increments of 0.025″ (1 × 0.025″ = 0.025″). The thimble scale divides each small barrel line increment into 25 parts, so each increment on the thimble is 0.001″. Looks like 0.018—almost 0.019″ but not quite. Let's again ignore the vernier scale for a minute.

Looking at the same drawing, the following addition of items gives the reading:

0.100″	—Large number barrel reading
+ 0.025″	—One increment on barrel
+ 0.018″	—Another increment on barrel
0.143″	—Total reading

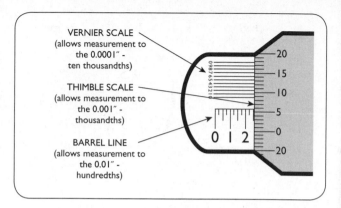

VERNIER SCALE
(allows measurement to
the 0.0001″ -
ten thousandths)

THIMBLE SCALE
(allows measurement to
the 0.001″ -
thousandths)

BARREL LINE
(allows measurement to
the 0.01″ -
hundredths)

Again, each large number on the barrel line indicates 0.100″ of space. The micrometer shown here has opened at least two of these numbers. The lines between each large number on the barrel are in increments of 0.025″, and we see two lines and more. In addition, the thimble shows five thousandths and a little more. Let's continue to ignore the vernier scale at this time.

Looking at the same drawing, the following addition of items gives the reading:

$$
\begin{array}{ll}
0.100'' & \text{—Large number barrel reading} \\
0.100'' & \text{—Second large number on barrel} \\
+\,0.025'' & \text{—One increment on barrel} \\
+\,0.025'' & \text{—Another increment on barrel} \\
+\,0.005'' & \text{—Increment on thimble} \\
\hline
0.255'' & \text{—Total reading (and a little more)}
\end{array}
$$

This reading is accurate to the thousandth. We have a value of a little more than 1/4″ (0.250 = 1/4″).

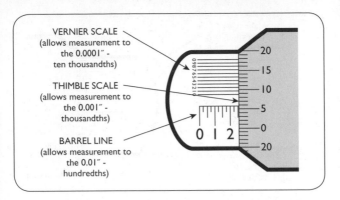

The last piece of the puzzle is to learn to read the vernier scale to achieve a measurement accurate to the ten–thousandth of an inch (0.0001")—the fourth decimal place. Look carefully at the vernier scale. What is the number on the line that *lines up closest* to any of the lines on the thimble? Sharp eyes will see it's the 8, or 0.0008". You read it as eight ten–thousandths of an inch. Adding this into the other measurements gives a final reading accurate to the ten–thousandth of an inch.

Looking at the same drawing, the following addition of items gives the reading:

$$
\begin{array}{rl}
0.100'' & \text{—Large number barrel reading} \\
0.100'' & \text{—Second large number on barrel} \\
+\ 0.025'' & \text{—One increment on barrel} \\
+\ 0.025'' & \text{—Another increment on barrel} \\
+\ 0.005'' & \text{—Increment on thimble} \\
+\ 0.0008'' & \text{—Increment on thimble} \\
\hline
0.2558'' & \text{—Total reading}
\end{array}
$$

This reading is accurate to the thousandth. We have a value of a little more than 1/4" (0.250 = 1/4").

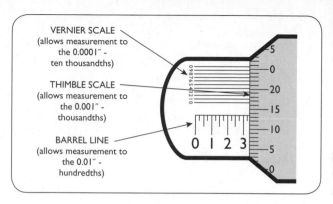

One more to make sure. Barrel—at least three big numbers (0.300″). Then one small increment (0.025″). Thimble—thirteen increments (0.013″). And then the vernier, the line that most closely lines up with a line on the thimble? Looks like the six (0.0006″). Add them up:

0.100″	—Large number barrel reading
+ 0.100″	—Second large number on barrel
+ 0.100″	—Third large number on barrel
+ 0.025″	—One increment on barrel
+ 0.013″	—Increment on thimble
+ 0.0006″	
0.3386″	—Total reading

Three more examples follow. As you work with micrometers it will become easier and you will find you no longer need pencil and paper to add the numbers. It's easier to add them in your head and write down the result.

If you are fitting a bearing to a shaft, the bearing dimensions and shaft dimensions are given to the ten–thousandth of an inch (0.0001″). Accuracy only to the thousandth (0.001″) is not good enough. If you are doing bearing work, you need a micrometer with the vernier scale to do the work properly.

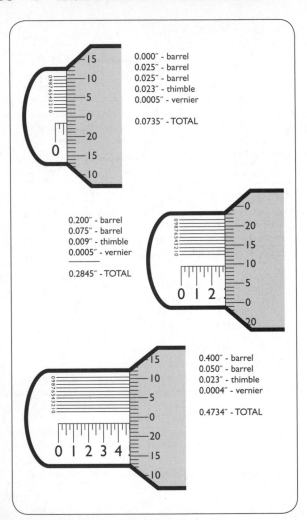

USING A STROBE LIGHT

The strobe light is an invaluable tool in troubleshooting problems that may be encountered in equipment maintenance. Usually the first use of a strobe is to determine the revolutions per minute (RPM) of machines—that is, their speed of operation. It is a fact that when the strobe light blinks the flash rate measured in cycles per minute (CPM) equal to the RPM of a rotating shaft or spinning element, the part in motion appears to be "dead–stopped" under the presence of the light. Many craftsmen do not realize that a spinning element also appears dead–stopped at other flash frequencies, which are submultiples of the actual RPM of the equipment. Sometimes, this phenomenon, from a branch of physics (optics), leads the craftsperson to erroneously associate the incorrect RPM with a component. There is a formula and procedure to ensure that this does not happen and that the observations lead to the correct RPM in every case.

First, three facts need to be known:

■ A flashing strobe light makes a rotating piece of equipment appear dead–stopped at true RPM and any submultiple of that RPM. Submultiples are 1/2, 1/3, 1/4. 1/5, 1/6, and so on, of the true RPM.

What are the first four submultiple flash rates if the true rotating speed of a fan has a value of 3600 RPM?
1/2 of 3600 = 1800 (first)
1/3 of 3600 = 1200 (second)
1/4 of 3600 = 900 (third)
1/5 of 3600 = 720 (fourth)
What is the next submultiple?
1/6 of 3600 = 600
Will the fan appear dead–stopped if the strobe rate is set to 450 flashes per minute?

Answer: 1/8 of 3600 = 450; hence, the answer is yes.

■ To make sure you are not seeing a "double" or "triple" under the strobe light, the determination of a dead–stopped condition is made using a mark or line that appears as a single radius or that can appear only once under the dead–stopped condition. This is called a *single mark*.

■ It is a fact that a single mark occurs only at a strobe flash rate of 1 × RPM and submultiples (1/2, 1/3, etc.) and will never appear at 2 × RPM, 3 × RPM, 4 × RPM, or higher. These higher flash rates are called *harmonics*.

ROTOR MUST BE MARKED WITH A RADIAL LINE OR HAS TO HAVE AN IDENTIFIER THAT CAN ONLY APPEAR ONE TIME

EXAMPLE #1: RADIAL MARK PAINTED ON ROTOR.

EXAMPLE #2: KEY AND KEYWAY CAN BE USED. CAN ONLY APPEAR ONCE.

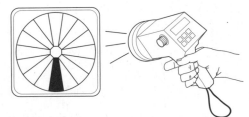

ONE BLADE ON FAN IS BLACKENED OR MARKED. THIS BLADE SHOULD ONLY APPEAR ONCE IN A "DEAD STOPPED" VIEW OF THE FAN USING THE STROBE.

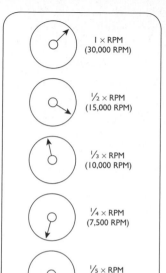

"Stopped" flash rates for a rotor whose true RPM = 30,000. Single pictures appear at submultiples of rotor speeds, $1/2$, $1/3$, $1/4$, etc. as shown.

Each of the individual pictures shows the rotor "stopped" at various flash rates from $1 \times$ RPM to as low as $1/6 \times$ RPM. The equation for determining the true speed is:

$$RPM = \frac{R_1 \times R_2}{R_1 - R_2}$$

R_1 and R_2 are any two adjacent flash rates in descending order. Hence, if R_1 was 30,000, then R_2 would be 15,000. If, however, R_1 was 6000, then R_2 would be 5000.

$$RPM = \frac{30K \times 15K}{30K - 15K} = 30,000$$

$$RPM = \frac{6K \times 5K}{6K - 5K} = 30,000$$

Using either of these sets of numbers for R_1 and R_2 and plugging them into the formula gives a true RPM value of 30,000 as a solution.

Notice that a strobe light does *not* have to be capable of firing at a rate that equals the true RPM of a machine in order to determine the speed. In fact, most strobe lights are not capable of firing at 30,000 flashes per minute, yet if other, lower values of R_1 and R_2 are obtained, the true speed can still be determined. It is also good form to obtain three readings in descending order to double–check your calculations.

Imagine a situation in which your strobe light cannot flash at 30,000, which is the true speed of the fan. You obtain three descending "single pictures," making sure you don't miss one. Your three flash rates are 10,000, 7500, and 6000. Using the equation and values of $R_1 = 10,000$ and $R_2 = 7500$ gives the following:

$$RPM = \frac{10,000 \times 7500}{10,000 - 7500} = \frac{175,000,000}{2500} = 30,000$$

And substituting new values in the formula yields:

$$RPM = \frac{7500 \times 6000}{7500 \times 6000} = \frac{45,000,000}{500} = 30,000$$

A belt–driven fan shows readings of 1200, 800, and 600. What is the true RPM of the fan?

Answer: See upside–down text.

$$RPM = \frac{1,200 \times 800}{1,200 - 800} = \frac{960,000}{400} = 2,400$$

$$RPM = \frac{800 \times 600}{800 - 600} = \frac{480,000}{200} = 2,400$$

Use of the strobe light to make "slow–motion" studies is a valuable troubleshooting tool. Turning the strobe to the exact same flash rate as the speed of the machine produces a stopped picture. Then slightly detuning the light makes the machine appear to spin in slow motion, allowing the observer to look for defects that might show up only in a spinning and loaded situation. Some things to look for using the slow–motion study include:

■ Keys moving back and forth in the keyseat
■ Cracks in welds on fan blades that pull apart from centrifugal forces while the fan is running and disappear when the fan stops and the crack closes
■ Blade wobble or improper tracking on a propeller or fan

- Visual inspection of a flexible coupling between a motor and a machine while the unit is running, allowing inspection without the need to shut the machine down
- Checking visible bearings to determine whether the outer race is creeping around in the bearing housing, which indicates a loose fit
- Movement of a setscrew up and down due to vibration or an untightened condition
- Observing the lettering on the top of running V–belts to determine the part numbers for reorder while the machine is running
- Checking multiple sets of V–belts to determine whether they are tracking together, which indicates good sheave alignment or a condition of misalignment due to sheaves operating at an angle to each other

VIBRATION MEASUREMENT

All rotating equipment (motors, fans, blowers, centrifugal pumps, turbines) vibrate. Vibration is just a force operating on the machine's bearings and other components that repeats itself again and again. In the United States the total movement of a machine (how far it shakes) is called *displacement* and is measured in *mils*. One mil equals 0.001″ of

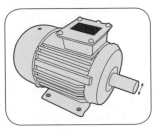

shake or displacement. Normally, mils are measured as peak to peak, or the total amount of movement that the case of the machine makes from side to side.

Looking in at the shaft end of a motor that is vibrating, you can easily see what is meant by "peak–to–peak displacement."

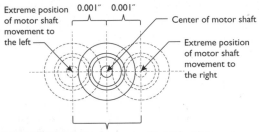

DISPLACEMENT of vibration measured in MILS peak to peak. Measurement is from the two extreme positions of the centerline of the motor shaft as shown.

> *What is the displacement peak to peak in the figure shown here?*
> *Answer: The movement to the right is 0.001″ and to the left it is*
> *0.001″. The displacement of this motor shaft is 0.002″ peak to*
> *peak, or a total displacement of 2 mils of vibration.*

The next term used in vibration measurement is an indication of how often a vibration repeats. The motor shaft, shown in the previous figure, moves from the center to the extreme right position and then back through the center to the extreme left position, and finally back to the center again. At that point, the motion repeats itself. This unique motion (center to right–back through center to left–back to center) is called a *cycle*. If you put your fingers on the motor case, you feel the motion again and again. Your fingers are sensing the force of vibration that is moving the motor side to side.

The measurement of how often a machine vibrates is called *frequency* and is measured in *cycles per minute* (CPM). While electrical workers measure frequency in CPS (cycles per second, or hertz), mechanical vibration is usually expressed in CPM. Conversion from CPS to CPM simply involves multiplying the CPS by 60 to obtain CPM (since there are 60 seconds in a minute).

$$CPM = CPS \times 60$$

> *If a vibration has a frequency of 30 cycles per second (30 hertz), what is the frequency in CPM?*
> *CPM = 30 × 60 = 1800 CPM*
>
> *If the frequency is 50 cycles per second, what is the CPM?*
> *CPM = 50 × 60 = 3000 CPM*
>
> *If the frequency of vibration is 3600 CPM, what is the frequency in CPS?*
>
> *CPS = CPM ÷ 60*
> *CPS = 3600 ÷ 60 = 60 cycles per second, or 60 hertz*

If displacement is graphed over a period of time, most mechanical rotating machinery has a sine–wave shape. The tops and bottoms of the waves represent the extreme movements. The total height from the bottom to the top is the peak–to–peak displacement. (**Heavy line** shows one cycle.)

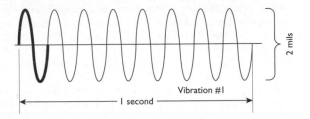

What is the frequency of Vibration #1 graphed here in CPM?
Count the cycles in 1 second and multiply by 60.

$$CPM = CPS \times 60$$

CPM = 8 cycles per second × 60 = 480 cycles per minute
What is the displacement?
Answer: 2 mils peak to peak.

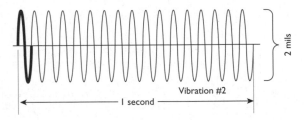

What is the frequency of Vibration #2 graphed here in CPM?
Count the cycles in 1 second and multiply by 60.

$$CPM = CPS \times 60$$

CPM = 17 cycles per second × 60 = 1020 cycles per minute
What is the displacement?
Answer: 2 mils peak to peak.

Vibration #1 and Vibration #2 have the same displacement but different frequencies. Vibration #3 (shown next) has the same frequency as Vibration #1 but different displacement.

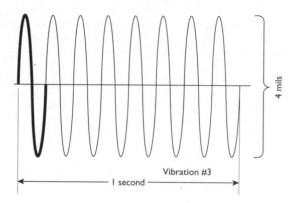

What is the frequency of Vibration #3 graphed here in CPM?
Count the cycles in 1 second and multiply by 60.

$$CPM = CPS \times 60$$
$$CPM = 8 \text{ cycles per second} \times 60 = 480 \text{ cycles per minute}$$

What is the displacement?
Answer: 4 mils peak to peak.

Vibration #2 and Vibration #3 are more *severe* than Vibration #1. Imagine each to be a roller–coaster ride. Vibration #1 would be fun, but #2 and #3 might snap your neck. Vibration #2 has more ups and downs in a minute, and #3 has higher lifts and lower drops. Both will cause severe injury to the riders.

Vibration has a similar effect on machinery and bearings. More displacement or higher frequency can cause damage.

The bearings in any machine are subjected to force as the machine does useful work. The machine designer or engineer sizes and chooses bearings that will provide for good service life by handling the forces that are considered to be normal and expected. Bearings are *not* designed to contend with destructive forces due to adverse machinery conditions or machine element faults such as unbalance, misalignment between components, improper belt tension, eccentricity, uneven wear, looseness, and even vibration caused by surrounding equipment.

In 1930, a noted engineer, Arvid Palmgren, published the famous failure–fatigue equation for the life of an antifriction bearing measured in revolutions. Palmgren noted that all antifriction bearings do fail eventually, but their service life is very quickly diminished as forces increase. In fact, a ball bearing's service life varies as the cube (a number multiplied by itself 3 times) of the force on the bearing. This means that increasing the force on a bearing by twice what is normal has the effect of lowering the bearing's life by 8 times: $2 \times 2 \times 2 = 8$. If the bearing has to contend with an increase above normal of three times, then the bearing life is diminished by 27 times: $3 \times 3 \times 3 = 27$. That's a huge chunk of valuable machine running time that is lost, and it can produce some costly repairs.

A quick method of judging the severity of the force level on a bearing is accomplished by measuring the vibration that is experienced by the bearing, usually on the case of the machine near the location of the bearing's internal position.

Doubling the force on a bearing decreases its life by 8 times

As we have seen in the roller–coaster example, force can be excessive when either the *displacement* (mils) of the vibration or the *frequency* (CPM) of the vibration increases. An increase in both is even more dangerous. In 1939, another engineer, Thomas Rathbone, invented a vibration severity chart that indicated machine "health" based on vibration measured at or near a bearing position.

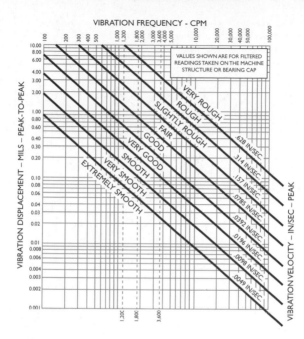

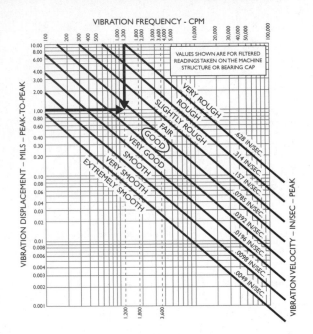

Assume a vibration of 1.0 mil at a frequency of 1200 CPM measured near the bearing housing on a particular machine. What is the severity of this vibration? The intersection of the two values falls in the "good" region of the chart shown in the figure. This level of vibration would show you that the machine has no high–level forces acting on it and that the service life of the machine should be longer than normal. Pumps, blowers, and fans would be okay.

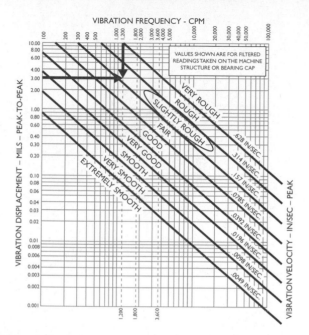

If the displacement of the vibration gets bigger while the frequency remains the same, this can result in a higher level of forces on the bearing. Look at the chart. There is a vibration of 3 mils with a frequency of 1200 CPM. It shows the same frequency as the previous example but a different displacement. Now the vibration is considered to be "slightly rough." The bearings are going to fail prematurely due to a higher level of forces on them.

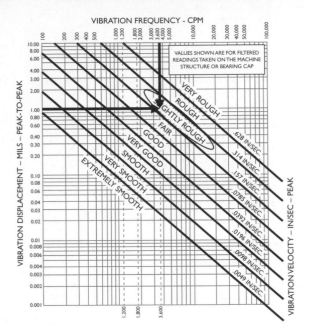

Conversely, if the frequency of the vibration increases but the displacement remains the same, this also results in a higher level of forces on the bearing. Look at the chart. There is a vibration of 1 mil with a frequency of 3600 CPM. It shows the same frequency as the first example but a different displacement. Again the vibration is in the "slightly rough" area. Bearing failure will again be expected prematurely.

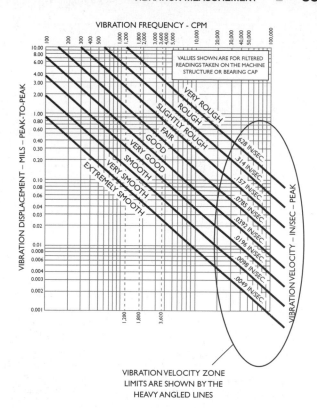

VIBRATION VELOCITY ZONE
LIMITS ARE SHOWN BY THE
HEAVY ANGLED LINES

There is a simpler way to make use of the chart. The heavy border lines that separate each zone of severity show the speed of vibration measured in inches per second (in./sec.). This is not the speed of the machine, but rather the speed at which the machine vibrates. It's analogous to the car having a speed of 60–MPH or 15–MPH, while the speed of the gasoline engine might not change. Using *vibration velocity (speed)* makes force measurement on a machine a one–number task.

Vibration Velocity (Force on a bearing)	Results of This Force on a Bearing	What Needs to Be Done?
< 0.15 in/sec	A less than normal amount of force. Bearing life will be excellent.	Spot check only if problems develop. Force levels are low.
0.15 in/sec	A normal amount of force. Bearing life should correspond with manufacturer's expectations.	Spot check only if problems develop. Force levels are normal.
0.30 in/sec	This is double the amount of force that would be expected. The bearing life is decreased by 8 times.	Pay attention. Attempt to troubleshoot to determine source of forces.
0.45 in/sec	Triple the amount of force expected. The bearing life is decreased by 27 times.	Take action. If source of problem has not been determined, intensify efforts.
0.60 in/sec	This level of force ruptures the supportive oil film necessary to lubricate a bearing. With both high forces and lubrication loss, bearing life is drastically decreased.	Consider shutting the machine down if the source of the forces can not be determined or corrected.
0.90 in/sec	Excessive force. Metal parts of bearing and surrounding supporting components (housing, shaft, etc.) will mushroom and deform.	Operation of the machine at this level of force is now ruining other parts. The cost to repair will be high.

The chart shown here, based on vibration velocity as measured on the housing of a machine at or near a bearing cap (antifriction bearing), can be used with good results to make decisions about what needs to be done.

It's interesting to use vibration to determine the exact cause of a high level of forces, but, in many cases, without training and expensive instrumentation it is difficult. It is more important to recognize that a problem exists and prepare to deal with it. Often, even without expensive instruments, a good tradesman can guess the cause of problems and take some simple steps to prove his theory true or false.

Shown here is a mechanical schematic of a horizontally driven pump and motor. The horizontal plane is chosen for taking vibration readings because there is usually more freedom to

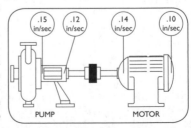

vibrate in that direction. Bolts and the force of gravity are locking down the pump in the vertical direction, and the bearings and piping are holding the pump in an axial direction. The horizontal direction is the "worst case" for vibration. Taking readings there will more rapidly identify problems.

The readings shown might be typical of a pump that has been installed properly. They show what might be considered normal for a pump and motor where the shafts have been aligned and the assembly has been anchored, shimmed, and grouted to the concrete. If there are no mechanical problems, such as misalignment, looseness, or unbalance, the forces on each bearing are normal, and the vibration readings prove it. Readings centering around values of 0.15 in/sec are very typical for most common machinery that is in good condition. This equipment should be left alone.

The higher the level of vibration, the closer to the source of the forces. Here is an example that demonstrates this fact. Notice the higher vibration measurements on each side of the coupling with suitable drop in vibration as readings are taken from bearing positions farther away. This is typical of poor alignment between the motor and the pump. The misalignment causes excessive vibration on both of the bearings nearest the coupling and will degrade them if permitted to continue. This type of check can be made on brand–new

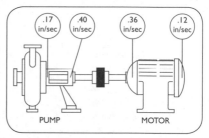

equipment to make sure the contractor does not leave you a problem by performing a less–than–high–quality alignment.

Although the details of machinery vibration analysis are outside the scope of this book, these two examples show the benefit of checking a piece of equipment using *velocity* standards to determine what is good and what is not.

SIMPLE VIBRATION SEVERITY TEST

While owning and using a vibration meter give a mechanic the power to determine many faults in a machine, there is an older, simple test to determine whether a vibration is harmful. It does not involve the use of an instrument. It's called the *nickel test*. It is very effective and produces good results.

Take a brand–new or "like new" nickel and find a place on the machine near a bearing cap that has a flat area. While the machine is running, carefully attempt to stand the nickel on its edge. Release your fingers slowly. If the nickel stands up and does not fall over, then the vibration level is low (usually below 0.15 in/sec). This instrument is usually in your pocket and it costs only five cents. This low–cost test can be used on most pumps, motors, fans, blowers, and other common pieces of rotating equipment. The nickel test might save you the time needed to run all the way back to the shop to grab that vibration instrument if it's really not needed. Remember to use a new nickel with a sharp edge.

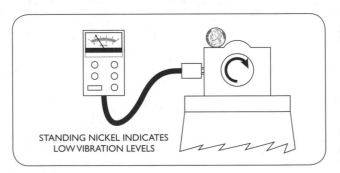

STANDING NICKEL INDICATES
LOW VIBRATION LEVELS

VIBRATION CALIBRATION

There is another field–worthy tip that you might like to try: Calibration of *yourself* to become sensitive to various vibration levels is possible. Using a vibration meter, find a bearing cap that measures 0.15-in/sec vibration velocity. Find another in your facility that measures 0.30 in/sec, and finally (if possible) find one that checks out at 0.45 in/sec or even 0.60 in/sec. After you read the instrument, safely place your fingers on the bearing cap to feel the vibration. Sometimes closing your eyes for a second or two just to focus on the feeling can help. It will not take long before you can touch a machine and provide a fairly well qualified guess about the vibration severity. You become the instrument.

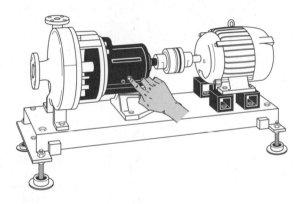

DETECTING A BAD BEARING

While there are many sophisticated electronic instruments that indicate bad antifriction bearings, the seasoned multi–craftsman can accomplish the same thing by listening to the sound of a bearing, with the help of some simple devices. A long screwdriver, a mechanic's stethoscope, or even the brim of a hard hat can be brought into play. The technique can be learned in less than a minute.

Good bearings exhibit a "beach sound." Imagine being at the seashore late at night and hearing the sounds of the waves lapping on the shore. That is the sound of a good bearing. Take a screwdriver and listen to several bearings in machinery that would be expected to be good. Start with fairly new equipment. Just a few good bearings will enable you to identify the beach sound. Once you've heard it, you are unlikely to forget it.

Bad bearings have a click, rumble, grinding, or other noise that indicates some level of internal damage. Basically a pothole (called a *spall*) is present in the inner raceway ball path, in the outer raceway ball path, or in one or more of the rolling elements themselves.

Some noises can be deceptive. On a centrifugal pump, an abnormal noise could be caused by a bad bearing or by cavitation occurring inside the pump casing. Momentarily stop the flow through the pump. If the noise disappears, it's cavitation and not the bearing. If the noise stays, then it's the bearing.

Ball bearings can likewise be deceptive. If a ball is spalled and the spall rolls into the race path, it makes noise. The ball can turn while it rolls so the spall is outside the path and the noise disappears. The bearing does not "heal" itself. In time, the noise will appear again when the ball changes its axis of rotation.

5

LUBRICATION

PREPACKING A BEARING WITH GREASE

When replacing bearings in a machine, it is necessary to prepack the new bearings prior to installing them. A bearing packer can be used to prepack bearings, or they can be pregreased using the *grease in palm* method.

The standard bearing-packing tool contains two convex-shaped plates attached to a threaded rod with a grease zerk fitting on the end of the threaded rod. With the bearings in place in the packing tool, pump grease into the zerk fitting, forcing grease through the rollers or balls of the bearing. This bearing is now ready for installation.

If no bearing-packing tool is available, suitable results can be obtained using the grease in palm method. Although a messy process, it is equally effective if done properly and carefully. As the name implies, grease

is placed in the palm of one hand and then the other hand methodically rolls and rotates the bearing into the grease, carefully forcing grease through all rollers or balls. After prepacking of both the inner and the outer bearings is completed, the bearing should be placed on a totally clean surface. Dirt in the greased bearing acts as an abrasive on the bearings and races and can drastically shorten bearing life.

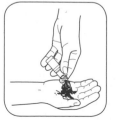

CALIBRATION OF A GREASE GUN

Many machines come with manuals that provide lubrication information.
Usually the amount of grease to be added is given in *ounces.*

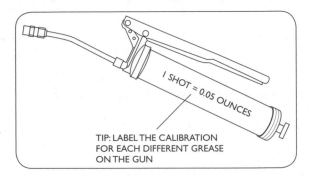

I SHOT = 0.05 OUNCES

TIP: LABEL THE CALIBRATION
FOR EACH DIFFERENT GREASE
ON THE GUN

This can present a problem, since grease guns do not dispense grease
by the ounce but by the *shot.* Pumping the grease gun with a full stroke
gives a metered amount of grease. Since a grease gun can be thought of
as a highly accurate positive displacement pump, it takes but a few steps
to get a pretty accurate measurement of the number of shots per ounce.

■ Weigh a small clean container on an accurate scale. The
scale should be accurate to 0.01 oz. Record the weight of
the container. If the scale has the ability to "tare" the container (make the container appear to weigh nothing), then
make sure of that feature. You want to weigh grease, not the
container.

■ Slowly dispense 10 shots of grease into the container. Work
the grease gun slowly, and be careful to make a full stroke
each time. *Tip: Don't do this to a grease gun on which you've
just replaced the grease cartridge. There is too much chance
of pumping a few air bubbles when you start. Use a gun from
which about 10 percent of the grease has already been used.*

Shake the gun after the tenth time to make sure all the grease gets into the container. You want to be accurate.

▪ Weigh the container with the grease, and subtract the original weight you recorded for just the container.

▪ The value you now have is the weight in ounces of 10 shots of grease. Divide this number by 10 to obtain the weight in ounces of 1 shot of grease. You can do this easily by moving the decimal point over one place to the left. The reason you dispensed 10 shots was to get an average number for each shot and make up of individual discrepancies that might take place on any given stroke of the handle of the grease gun.

▪ This is the number that you will divide into the number of ounces given in the maintenance manual to obtain the number of shots of grease needed when the equipment requires lubrication.

A glass beaker is used to catch the grease. The beaker weighs 1.5 oz. Ten shots of grease were added to the beaker. The new weight is 2.0 oz. The maintenance manual calls for 3/4 oz of grease to be added when relubrication is required. What is the number of shots required?

10 shots of grease and beaker	*= 2.0 oz*
Beaker alone	*= −1.5 oz*
Weight of 10 shots of grease	*= 0.5 oz*

Moving decimal to the left:

$$0.5 \ oz \div 10 \ shots = 0.05 \ oz/shot$$

Divide 3/4 ounce (0.750 ounce) by the ounces per shot:

$$0.750 \div 0.05 = 15 \ shots$$

For 7/8 ounce, how many shots are required?
Answer: 17.5 shots.
What if 10 shots weighed 0.04 oz and 5/8 (0.625) oz was called for?
  Answer: 15.6, rounded to 16 shots.

HOW MUCH GREASE TO ADD TO A BEARING

If you don't have the manufacturer's manual showing the correct amount of lubricant to add to a bearing, the bearing companies provide an easy-to-use formula to calculate the amount in ounces or grams. The two formulas require knowledge of the outside diameter (OD) of the bearing as well as the width of the bearing measured either in inches or millimeters.

$$G_g = 0.005 \times D_{mm} \times B_{mm} \text{ grams}$$

G_g = grease quantity (in grams)
D_{mm} = Bearing outside diameter (OD) in millimeters
B_{mm} = Bearing width in millimeters

or

$$G_o = 0.114 \times D_{in} \times B_{in} \text{ ounces}$$

G_o = grease quantity (in ounces)
D_{in} = Bearing outside diameter (OD) in inches
B_{in} = Bearing width in inches

Keep in mind that these calculations give the suggested quantity of grease to add during a *regreasing* of a bearing as part of your lubrication preventive maintenance (PM) program. The quantity is *not* the amount of grease used to pack a bearing when it is first installed. There is a difference.

Consider a medium-duty ball bearing with a 25-mm bore. That would be a 305 bearing (300 series is medium duty, and $05 \times 5 = 25$ mm—the bearing bore).

Using the bearing dimensional table shown in this Mini-Ref or from any reliable source, we can see that the OD of the bearing is 2.4409″ and the width is 0.6693″. This provides enough information to determine the regreasing quantity needed to replenish the correct amount of grease during the PM cycle.

Inserting these values into the formula for ounces yields the following:

$$G_o = 0.114 \times D_{in} \times B_{in} \text{ oz}$$

$$G_o = 0.114 \times 2.4409 \times 0.6693$$

$$G_o = 0.186 \text{ oz}$$

If the grease gun dispenses 0.05 oz per shot, then G_o (0.186) is divided by 0.05 to give 3.72 shots, rounded to 4 shots, each time the bearing is lubricated on the PM program.

How much grease is required for a relubrication of a 419 ball bearing? The grease gun has been calibrated and gives 0.05 oz per shot.
From the bearing dimensional tables in this Mini-Ref:
  OD of bearing = D_{in} = 9.8425"
  Width of bearing = B_{in} = 2.1654"

$$G_o = 0.114 \times D_{in} \times B_{in} \text{ oz}$$

$$G_o = 0.114 \times 9.8425 \times 2.1654$$

$$G_o = 2.429 \text{ oz}$$

$$2.429 \div 0.05 \text{ oz per shot} = 48.58 \text{ shots,}$$

$$\text{rounded to 49 shots}$$

How much grease is required for a relubrication of a 220 ball bearing? The grease gun dispenses 0.45 oz with 10 shots.
Answer: 24.03 shots, rounded to 24 shots.

GREASE RELUBRICATION INTERVAL

The amount of time required before relubrication of an antifriction bearing is dependent on three things: the ID of the bearing, the speed (RPM) of the bearing, and the type of bearing (ball, roller, cylindrical, etc.). The amount of time can be determined by a relubrication chart.

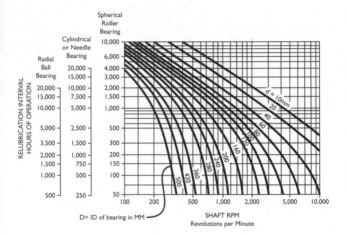

Keep in mind that this chart gives the number of hours between relubrications for a machine that is considered to be in operation 24/7, that is, continuous duty. A week of continuous duty consists of 24 × 7 = 168 hours. A continuous year is 24 × 365 = 8760 hours. A month of continuous duty would be approximately 8760 divided by 12 months = 730 hours.

If a machine operated only 8 hours a day for a year, the actual lubrication interval would be increased by three times, since the machine is running only one-third of the time available in a year. Example: A machine shows about a 4-month interval on the chart but 4 × 3 = 12 months, so the actual time would be yearly.

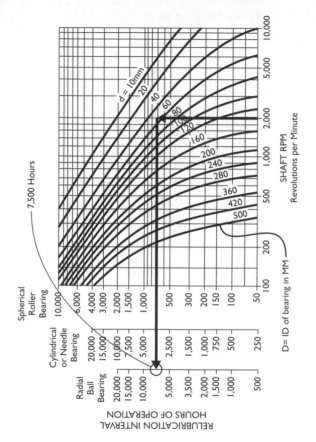

Imagine a pump that uses grease fittings and has *ball bearings* that are *60 mm* in inside diameter (ID) and a speed of *1750 RPM*. It operates 24/7. This is a fairly common situation throughout the country.

Using the chart, find the RPM on the bottom axis and draw a straight line up until it intersects the correct ID of the bearing. Now project a line to the left until it crosses the proper type of bearing, and read the hours of operation until regreasing is required. The answer is 7000 hours. If there are 168 hours in a week, this pump would require regreasing every $7500 \div 168 = 45$ weeks, or a little over 11 months. Lubrication is similar to horseshoes in that close counts. Set your lubrication regreasing at about once per year, and you will be just fine.

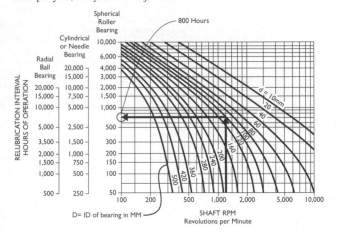

Another example is a pillow block bearing for a large fan that uses *spherical roller* bearings. The ID of the bearings is *100 mm* and the speed of rotation is *1150 RPM*.

On the bottom axis locate the point for 1150 RPM and draw a line up to the 100-mm curve. Stop there, and project a line to the left until it intersects the time line for a spherical roller bearing. The number of hours is 800. Since a month contains 730 hours, this bearing needs to be relubricated monthly. Again, that's close enough for lubrication. You don't need to be dead on the date, just close.

The next page presents a larger version of the chart that may be used to determine the relubricaton frequency for your own machinery or to give answers to the practice examples that follow in the Mini-Ref.

GREASE RELUBRICATION INTERVAL

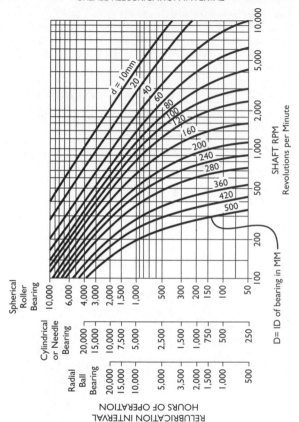

Your plant has two pumps that require greasing. The maintenance manual has been lost, misplaced, or relocated. A friendly argument develops between you and another mechanic about how much and how often the bearings should be lubricated. He thinks that 15 shots of grease every 3 months is about right, and you believe that is way too much. The pump uses some 208 ball bearings. The motor powering the pump shows a nameplate rating of 3550 RPM and it drives the pump with a flexible coupling. Your grease gun has been calibrated at 0.05 oz per shot. Who is right?

$$G_o = 0.114 \times D_{in} \times B_{in} \text{ oz}$$
$$G_o = 0.114 \times 3.1496 \times 0.7087 = 0.254 \text{ oz}$$
$$0.254 \text{ oz} \div 0.05 \text{ shot per oz} = 5.08, \text{ or}$$
$$5 \text{ shots}$$

Using the chart on the previous page and values of 3550 RPM for speed and 40 mm for the ID of the bearing, the relubrication interval for a ball bearing is about 4500 hours.

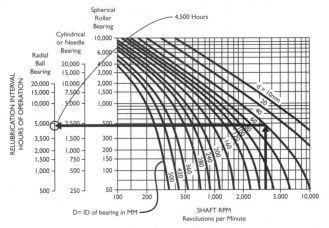

No contest! You win—hands down!

6

BEARINGS

BEARING IDENTIFICATION NUMBERS

It is not unusual for manufacturers of mechanical equipment to post information about the bearings on the nameplate affixed to the machinery. There are two conventions that are popular.

The first shows a listing of the basic bearing number and the series number and might also include the name of the manufacturer and the code for that manufacturer's type of bearing.

SKF — 6 2 20 XX

Hardware

Indication if bearing is open, shielded, or sealed. See chart on next page

Manufacturer

SKF
Fafnir
BCA
Hoover
MRC
New Departure
FAG
Others

Basic Number

This number (two digit) x 5 = Bore of the bearing in millimeters (20 x 5 = 100 mm for this bearing). Exceptions are:
00 = 10 mm
01 = 12 mm
02 = 15 mm
03 = 17 mm
All others follow the rule

Type

Unique to the manufacturer. Normally a one digit or two digit number is shown here. One of the most commonly used digits by many manufacturers is "6" which indicates a single row, radial, deep groove, ball bearing (called a CONRAD). A number of "7" often indicates single row, angular contact ball bearing rated for thrust. Manufacturer's catalogs need to be consulted for other numbers that are in use.

Series Number

100 = Extra-Light Duty
200 = Light Duty
300 = Medium Duty
400 = Heavy Duty

<u>SKF—6 2 20 XX C3</u>

Hardware designated by the XX position in the bearing number usually indicates whether the bearing has shields or seals and how many. Here is a chart showing hardware markings for some leading manufacturers.

	One Shield	Two Shields	One Seal	Two Seals	Seal and Shield	Open with Snap Ring
BCA Suffix	S	SS	D	DD	DS	L
Fafnir Suffix	D	DD	P	PP	PD	G
FAG Suffix	Z	2Z	RS	2RS	RSZ	NR
Federal Suffix	F	FF	R	RR	RF	CG
Hoover Suffix Prefix	– 7	– 77	– 9	– 99	– 97	G –
MRC Suffix	F	FF	Z	ZZ	ZF	G
ND Prefix	7	77	9	99	97	4
N-H Suffix Prefix	P11 –	PP11 –	K –	KK –	KP –	– 4
SKF Suffix	Z	2Z	RS	2RS	RSZ	NR

<u>SKF—6 2 20 XX C3</u>

At the very end of the bearing number, most manufacturers include a value for the internal clearance inside the bearing itself. This clearance can be thought of as the distance between a ball and the outer race, which can be closed a bit when a bearing is press-fitted onto a shaft, thereby expanding the inner race. The preceding example shows the clearance number as "C3."

Bearing Clearance Number Chart		
Clearance Number	**Indication of Space in Bearing**	**Might Be Found on the Following**
C1	Less than C2	Precision machinery, e.g., a lathe or milling machine
C2	Less than normal	Semiprecision machinery, e.g., a drill press
Null (nothing shows)	Normal clearance	Most rotating equipment
C3	Greater than normal	Electric motors
C4	Greater than C3	Higher-temperature machinery
C5	Greater than C4	Very high temperature machinery

0000 A A A 00 A AA A 0 0

Bore

Bearing bore in millimeters
(up to 4 numbers)

Type—General

B = Radial ball
R = Cylindrical roller
S = Self aligning
(spherical roller)
T = Thrust, ball, or roller

Type–Specific to Ball Bearing

C = Single row, deep groove
(CONRAD)
L = Single row, filling slot
(Maximum type)
N = Single row, angular cont–
act, light thrust (15°)
A = Single row, angular con–
tact, medium thrust (25°)
T = Single row, angular con=
tact, heavy thrust (35° to
40°)

Type—Other

Used to further define bearing
types not normally used in motor
design

Duty Rating

02 = Light duty
03 = Medium duty
04 = Heavy duty

Tolerance

Defines specific
precision grades
—not normally
used in motor
design

Internal Clearance

2 = Less than normal
0 = Normal
3 = Greater than
normal

Special

Used to identify snap ring
modifications—not nor-
mally used in motor
design

Seats/Shields

E = Impervious seal, permanent
D = Impervious seal, removable
P = Shield, permanent
A = Shield, removable
Note: One letter used for each
side, i.e., PP = Double shielded
bearing

Seats/Shields

J = Steel, sheet stock, centered by
rolling elements
Y = Nonferrous, sheet stock,
centered by rolling elements
M = Bronze or brass machine, centered by rolling elements
T = Nonmetallic, centered by
rolling elements
X = Any cage type is acceptable

A piece of rotating machinery shows two bearings listed on the nameplate. The one nearly opposite the drive end is listed as SKF-6207-2Z, and the bearing near the drive end is stated as SKF-7307. What can be determined about the bearings from the numbers?

SKF-6207-2Z is manufactured by the SKF Bearing Company. It is a single-row, radial, deep-groove ball bearing (6) with double shields (2Z), and the bore of the bearing is 35 mm (7 × 5). It is a light-duty (200 series) bearing.

SKF-7308-NR is manufactured by the SKF Bearing Company. It is a single-row, angular-contact ball bearing (7) with open construction and a snap ring (NR). The bore of the bearing is 40 mm (8 × 5). It is a medium-duty (300 series) bearing.

Another piece of rotating equipment has two bearings listed on the nameplate. One is 70BC03JPP3 and the other is 110BT02M. What is known about these bearings?

70BC03JPP3	**110BT02M**
70 = 70-mm bore	110 = 110-mm bore
BC = Radial ball, single row, deep groove	BT = Radial ball, single row, angular contact
03 = Medium duty	02 = Light duty
J = Steel cage	M = Bronze cage
PP = Double shield	
3 = C3, larger-than-normal internal clearance	

BEARING DIMENSION TABLES

Bearing Dimensions—100 Series, Extra Light						
Bearing Number	**Bore (Inside Diameter)**		**Outside Diameter**		**Width**	
	Millimeters	**Inches**	**Millimeters**	**Inches**	**Millimeters**	**Inches**
100	10	0.3937	26	1.0236	8	0.3150
101	12	0.4724	28	1.1024	9	0.3150
102	15	0.5906	32	1.2598	9	0.3543
103	17	0.6693	35	1.3780	10	0.3937
104	20	0.7874	42	1.6535	12	0.4724
105	25	0.9843	47	1.8504	12	0.4724
106	30	1.1811	55	2.1654	13	0.5118
107	35	1.3780	62	2.4409	14	0.5512
108	40	1.5748	68	2.6772	15	0.5906
109	45	1.7717	75	2.9528	16	0.6299
110	50	1.9685	80	3.1496	16	0.6299
111	55	2.1654	90	3.5433	18	0.7087
112	60	2.3622	95	3.7402	18	0.7087
113	65	2.5591	100	3.9370	18	0.7087
114	70	2.7559	110	4.3307	20	0.7874
115	75	2.9528	115	4.5276	20	0.7874
116	80	3.1496	125	4.9213	22	0.8661
117	85	3.3465	130	5.1181	22	0.8661
118	90	3.5433	140	5.5118	24	0.9449
119	95	3.7402	145	5.7087	24	0.9449
120	100	3.9370	150	5.9055	24	0.9449
121	105	4.1339	160	6.2992	26	1.0236

Note: Dimensions found in these tables are useful in the formulas for regreasing amounts and regreasing intervals for antifriction bearings discussed elsewhere in this Mini-Ref.

Ball Bearing Dimensions—200 Series, Light Duty						
Bearing Number	**Bore (Inside Diameter)**		**Outside Diameter**		**Width**	
	Millimeters	Inches	Millimeters	Inches	Millimeters	Inches
200	10	0.3937	30	1.811	9	0.3543
201	12	0.4724	32	1.2598	10	0.3937
202	15	0.5906	35	1.3780	11	0.4331
203	17	0.6693	40	1.5748	12	0.4724
204	20	0.7874	47	1.8504	14	0.5512
205	25	0.9843	52	2.0472	15	0.5906
206	30	1.1811	62	2.4409	16	0.6299
207	35	1.3780	72	2.8346	17	0.6653
208	40	1.5748	80	3.1496	18	0.7087
209	45	1.7717	85	3.3465	19	0.7480
210	50	1.9685	90	3.5433	20	0.7874
211	55	2.1654	100	3.9370	21	0.8268
212	60	2.3622	110	4.3307	22	0.8661
213	65	2.5591	120	4.7244	23	0.9055
214	70	2.7559	125	4.9213	24	0.9449
215	75	2.9528	130	5.1181	25	0.9843
216	80	3.1496	140	5.5118	26	1.0236
217	85	3.3465	150	5.9055	28	1.1024
218	90	3.5433	160	6.2992	30	1.1811
219	95	3.7402	170	6.6929	32	1.2598
220	100	3.9370	180	7.0866	34	1.3386
221	105	4.1339	190	7.4803	36	1.4137
222	110	4.3307	200	7.8740	38	1.4962

Note: Dimensions found in these tables are useful in the formulas for regreasing amounts and regreasing intervals for antifriction bearings found elsewhere in this Mini-Ref.

Ball Bearing Dimensions—300 Series, Medium						
Bearing	Bore (Inside Diameter)		Outside Diameter		Width	
Number	Millimeters	Inches	Millimeters	Inches	Millimeters	Inches
300	10	0.3937	35	1.3780	11	0.4331
301	12	0.4724	37	1.4567	12	0.4724
302	15	0.5906	42	1.6535	13	0.5118
303	17	0.6693	47	1.8504	14	0.5512
304	20	0.7874	52	2.0272	15	0.5906
305	25	0.9843	62	2.4409	17	0.6693
306	30	1.1811	72	2.8346	19	0.7480
307	35	1.3780	80	3.1496	21	0.8268
308	40	1.5748	90	3.5433	23	0.9055
309	45	1.7717	100	3.9370	25	0.9843
310	50	1.9685	110	4.3307	27	1.0630
311	55	2.1654	120	4.7244	29	1.1417
312	60	2.3622	130	5.1180	31	1.2205
313	65	2.5591	140	5.5118	33	1.2992
314	70	2.7559	150	5.9066	35	1.3780
315	75	2.9528	160	6.2992	37	1.4567
316	80	3.1496	170	6.6929	39	1.5354
317	85	3.3465	180	7.0866	41	1.6142
318	90	3.5433	190	7.4803	43	1.6929
319	95	3.7402	200	7.8740	45	1.7717
320	100	3.9370	215	8.4646	47	1.8504
321	105	4.1339	225	8.8583	49	1.9291
322	110	4.3307	240	9.4480	50	1.9685
324	120	4.7244	260	10.2362	55	2.1654
326	130	5.1181	280	11.0236	58	2.2835
328	140	5.5118	300	11.8110	62	2.4409
330	150	5.9055	320	12.5984	65	2.5591
332	160	6.2992	340	13.3858	68	2.6772
334	170	6.6929	360	14.1732	72	2.8346
336	180	7.0866	380	14.9606	75	2.9528
338	190	7.4803	400	15.7480	78	3.0709
340	200	7.8740	420	16.8354	80	3.1496
342	210	8.2677	440	17.3228	84	3.3031
344	220	8.6614	460	18.1002	88	3.4646
348	240	9.4488	500	19.6850	95	3.7402
352	260	10.2362	540	21.2598	102	4.0157
356	280	11.0236	580	22.8346	108	4.2520

Ball Bearing Dimensions—400 Series, Heavy Duty					
Bearing Number	Bore (Inside Diameter)		Outside Diameter		Width	
	Millimeters	Inches	Millimeters	Inches	Millimeters	Inches
403	17	0.6693	62	2.4409	17	0.6693
404	20	0.7874	72	2.8345	19	0.7480
405	25	0.9843	80	3.1496	21	0.8268
406	30	1.1811	90	3.5433	23	0.9055
407	35	1.3780	100	3.9370	25	0.9843
408	40	1.5748	110	4.3307	27	1.0630
409	45	1.7717	120	4.7244	29	1.1417
410	50	1.9685	130	5.1181	31	1.2205
411	55	2.1654	140	5.5118	33	1.2992
412	60	2.3622	150	5.9055	35	1.3780
413	65	2.5591	160	6.2992	37	1.4567
414	70	2.7559	180	7.0866	42	1.6535
415	75	2.9528	190	7.4803	45	1.7717
416	80	3.1496	200	7.8740	48	1.8898
417	85	3.3465	210	8.2677	52	2.0472
418	90	3.5433	225	8.8533	54	2.1260
419	95	3.7402	250	9.8425	55	2.1654
420	100	3.9370	265	10.4331	60	2.3622
421	105	4.1339	290	11.4173	65	2.5591
422	110	4.3307	320	12.5985	70	2.5759

Note: Dimensions found in these tables are useful in the formulas for regreasing amounts and regreasing intervals for antifriction bearings discussed elsewhere in this book.

BEARING INSTALLATION

Many common pieces of mechanical equipment, such as motors and pumps, make use of cylindrical (straight) bore bearings. The shaft is slightly larger than the bearing bore, and a press or interference fit is required to assemble the bearing to the shaft.

The methods used for installation are:

Cold installation:
Use a hammer and drive sleeve, arbor press, or hydraulic press for bearings with an outside diameter (OD) of 4″ or less.

Hot installation:
Use a hot oil bath, heating oven, cone heater, or induction heater for bearings with an OD of greater than 4″.

PRESS RAM

SHAFT

BEARING

SUPPORTS
FOR INNER
RACE

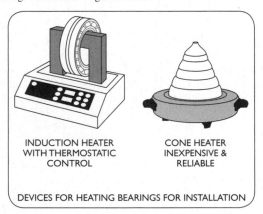

INDUCTION HEATER
WITH THERMOSTATIC
CONTROL

CONE HEATER
INEXPENSIVE &
RELIABLE

DEVICES FOR HEATING BEARINGS FOR INSTALLATION

Bearings that are cold-mounted should have the force applied only to the inner ring during assembly to the shaft. Pressing on the outer ring causes marks called *brinelling* to form on the ball paths and produce a noise in operation, as well as greatly shortening the life of the bearing. While cold-mounting *can* be done on bearings that are 4″ or less OD, thermal mounting can also be used. It's only when a bearing gets to be larger than 4″ OD that cold-mounting is no longer allowed.

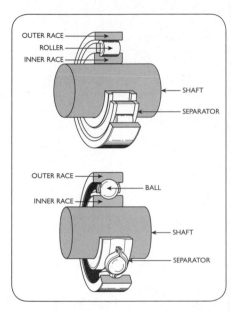

When heating a bearing to shrink-fit it to a shaft, a bearing temperature increase of 150°F above the shaft temperature provides sufficient expansion for mounting. Keep in mind that as it cools, the bearing contracts and grips the shaft tightly. The bearing must be heated uniformly with regulated heat and never heated above 250°F. Heating above this temperature degrades the properties of the bearing metal and softens the race paths. For bearings that have shields or seals, 210°F is advised as the maximum for thermal fitting, since temperatures in excess of this value cause breakdown of the grease or seal material. Actually, the boiling point of water (at sea level) is 212°F, and in an emergency a bearing can be placed in a watertight plastic bag and dropped into a reservoir of boiling water for 15 minutes to bring it to a uniform temperature for mounting. An inverted metal funnel sitting over a 100-watt incandescent bulb can also be used as a makeshift bearing heater. Care must be taken to keep a check on the bearing temperature using an infrared (IR) thermometer or temperature crayon so overheating is not present. Even a kitchen oven may be used to heat a bearing if the mechanic makes sure to place the bearing on the middle rack, set the preheat to 225°F, and place a cookie sheet on the racks above and below the bearing to keep the radiant coils of the oven from increasing the skin temperature of the bearing excessively due to the transfer of radiant energy.

When using heat to install a bearing, it is very important to keep a light but uniform pressure on the bearing to hold it against the shaft shoulder step in the shaft. Failure to keep pressure on the bearing results in the inner race pulling away from the shaft shoulder, causing the bearing to cock or misalign. Holding pressure for 5 minutes is about right for most bearings.

It might be a good time to point out that this simple check can be made on machinery while disassembly is

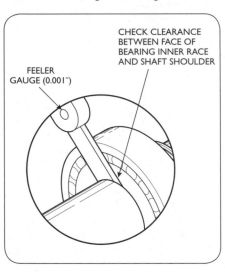

CHECK CLEARANCE BETWEEN FACE OF BEARING INNER RACE AND SHAFT SHOULDER

FEELER GAUGE (0.001″)

taking place. Finding excessive clearance between a bearing and the shoulder might be the answer to why the bearing failed prematurely or gave evidence of high vibration or excessive heat in operation.

FITTING OF ANTIFRICTION BEARINGS

In order for antifriction bearings to work well they must be round. Bearing manufacturers have spent years perfecting machining and grinding techniques to produce bearings that are round and stay round for years, assuming proper storage.

Ball or roller bearings have extremely accurate component parts that fit together with very close tolerances. The inner diameter (ID) of the inner ring and the OD of the outer ring are made to within close limits to fit their respective supporting members—the shaft and the housing. It makes sense that the shaft and the housing must also be machined to close limits so the fit of the bearing to these parts is maintained.

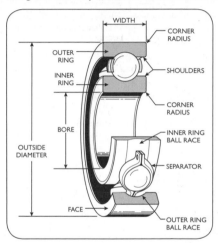

The inner and outer race rings are just that—rings. A ring is subject to deflection if it does not receive proper support. If the shaft that fits the bearing is "eggy" or "tapered," the bearing's inner race will take on the shape of the shaft. It will no longer be round. If the housing fit pinches the bearing's outer race, the outer race will distort and again the desired round raceway path is gone.

For the most part, on common machinery such as motors, pumps, fans, and blowers, the inner race of the bearing is an interference fit onto the shaft seat, which involves pressing the bearing onto the shaft or heating the bearing to expand the bore, installing it on the shaft, and then letting it cool down to lock itself in place. This process is called *fitting the bearing to the shaft.*

The fit of the outer race to the housing for similar machinery is usually a *tap* fit or a *light sliding* fit. A tap fit can mean that the OD of the bearing and the ID of the housing are the same dimension—the bearing will just fit if it is tapped on. A light sliding fit means that there is a tiny bit of clearance between the two parts and they should slide together without using any level of mechanical force.

There are two basic ways of installing a bearing onto a shaft: mechanical mounting and thermal mounting. Mechanical mounting may be done on ball bearings whose outside diameter is less than 100 mm (4″), on cylindrical bearings where the bore is less than 70 mm, and on tapered bearings where the bore is less than 250 mm. Thermal mounting is the acceptable method when a bearing has a 100-mm (4″) or greater OD. A point must be made here: While mechanical mounting is acceptable for bearings less than 4″ OD, there is nothing wrong with using thermal methods on those bearings.

Take a minute to check the table on the next page for a 320 bearing. Note the dimensions of the bearing bore as well as the shaft diameter required at the bearing seat. A portion of the table is shown here:

Basic Bearing Number	Nominal Bearing Bore	Bearing Bore Specification		Shaft Seat Specification	
		Max.	Min.	Max.	Min.
19	95	3.7402	3.7394	3.7409	3.7403
20	100	3.9370	3.9362	3.9377	3.9371
21	105	4.1339	4.1331	4.1350	4.1344

The bore of a 100-mm bearing can range from 3.9362″ minimum to 3.9370″ maximum, while the shaft must measure from 3.9371″ to 3.9377″. Looking closely shows that in every case the bearing is an interference or press fit onto the shaft.

BEARING BORE AND SHAFT SEAT DIMENSIONS

Basic Bearing Number	Nominal Bearing Bore	Bearing Bore Specification		Shaft Seat Specification	
		Max.	Min.	Max.	Min.
00	10	0.3937	0.3934	0.3939	0.3936
01	12	0.4724	0.4721	0.4726	0.4723
02	15	0.5906	0.5903	0.5908	0.5905
03	17	0.6693	0.6690	0.8895	0.6692
04	20	0.7874	0.7870	0.7879	0.7875
05	25	0.9843	0.9839	0.9848	0.9844
06	30	1.1811	1.1807	1.1816	1.1812
07	35	1.3780	1.3785	1.3785	1.3781
08	40	1.5748	1.5743	1.5753	1.5749
09	45	1.7717	1.7712	1.7722	1.7718
10	50	1.9685	1.9680	1.9690	1.9686
11	55	2.1654	2.1648	2.1660	2.1655
12	60	2.3622	2.3618	2.2628	2.2623
13	65	2.5591	2.5585	2.5597	2.5592
14	70	2.7559	2.7553	2.7565	2.7560
15	75	2.9528	2.9522	2.9534	2.9529
16	80	3.1496	3.1490	3.1502	3.1497
17	85	3.3465	3.3467	3.3472	3.3466
18	90	3.5433	3.5425	3.5440	3.5434
19	95	3.7402	3.7394	3.7409	3.7403
20	100	3.9370	3.9362	3.9377	3.9371
21	105	4.1339	4.1331	4.1350	4.1344
22	110	4.3307	4.3299	4.3318	4.3312
23	115	4.5276	4.5268	4.5287	4.5281
24	120	4.7244	4.7236	4.7255	4.7249
25	125	4.9213	4.9203	4.9226	4.9219
26	130	5.1181	5.1171	5.1194	5.1187
28	140	5.5118	5.5018	5.5131	5.5124
30	150	5.9055	5.9045	5.9071	5.9061
32	160	6.2992	6.2982	6.3008	6.3998
34	170	6.6929	6.6919	6.6945	6.6935
36	180	7.0856	7.0882	7.0882	7.0872
38	190	7.4803	7.4791	7.4821	7.4810
40	200	7.8740	7.8728	7.8758	7.8747
44	220	8.6614	8.6602	8.6632	8.6621
48	240	9.4488	9.4476	9.4506	9.4495
52	260	10.2362	10.2348	10.2382	10.2370
56	280	11.0236	11.0222	11.0256	11.0244

BEARING OD AND HOUSING DIMENSIONS—100 SERIES

Basic Bearing Number	Nominal Bearing OD	Bearing OD Specification		Housing Bore Specification	
		Max.	Min.	Max.	Min.
00	–	–	–	–	–
01	–	–	–	–	–
02	–	–	–	–	–
03	–	–	–	–	–
04	–	–	–	–	–
05	–	–	–	–	–
06	–	–	–	–	–
07	–	–	–	–	–
08	–	–	–	–	–
09	–	–	–	–	–
10	–	–	–	–	–
11	–	–	–	–	–
12	–	–	–	–	–
13	–	–	–	–	–
14	–	–	–	–	–
15	–	–	–	–	–
16	–	–	–	–	–
17	–	–	–	–	–
18	–	–	–	–	–
19	–	–	–	–	–
20	–	–	–	–	–
21	–	–	–	–	–
22	180	7.0866	7.0856	7.0882	7.0866
23	–	–	–	–	–
24	200	7.8440	7.8728	7.8758	7.8740
25	–	–	–	–	–
26	210	8.2677	8.2665	8.2695	8.2677
28	225	8.8583	8.8571	8.8601	8.8583
30	250	9.8525	9.8413	9.8443	9.8425
32	270	10.6299	10.6285	10.6319	10.6299
34	280	11.0236	11.0222	11.0256	11.0236
36	300	11.8110	11.8096	11.8130	11.8110
38	320	12.5984	12.5968	12.6006	12.5984
40	340	13.3858	13.3832	13.3880	13.3858

BEARING OD AND HOUSING DIMENSIONS—200 SERIES

Basic Bearing Number	Nominal Bearing OD	Bearing OD Specification		Housing Bore Specification	
		Max.	Min.	Max.	Min.
00	30	1.1811	1.1807	1.1819	1.1811
01	32	1.2598	1.2593	1.2608	1.2598
02	35	1.3980	1.3775	1.3790	1.3980
03	40	1.5748	1.5743	1.5758	1.5748
04	47	1.8504	1.8499	1.8514	1.8504
05	52	2.0472	2.0467	2.0484	2.0472
06	62	2.4409	2.4404	2.4421	2.4409
07	72	2.8346	2.8341	2.8358	2.8346
08	80	3.1496	3.1491	3.1508	3.1496
09	85	3.3465	3.3459	3.3479	3.3465
10	90	3.5433	3.5427	3.5447	3.5433
11	100	3.9370	3.9364	3.9384	3.9370
12	110	4.3307	4.3301	4.3321	4.3307
13	120	4.7244	4.7238	4.7258	4.7244
14	125	4.9213	4.9205	4.9229	4.9213
15	130	5.1181	5.1173	5.1197	5.1181
16	140	5.5118	5.5110	5.5134	5.5118
17	150	5.9055	5.9047	5.9071	5.9055
18	160	6.2992	6.2983	6.3008	6.2992
19	170	6.6929	6.6919	6.6945	6.6929
20	180	7.0866	7.0856	7.0882	7.0866
21	190	7.4803	7.4791	7.4821	7.4803
22	200	7.8740	7.8728	7.8758	7.8740
23	–	–	–	–	–
24	215	8.4646	8.4634	8.4664	8.4646
25	–	–	–	–	–
26	230	9.0551	9.0539	9.0569	9.0551
28	250	9.8425	9.8413	9.8443	9.8425
30	270	10.6299	10.6285	10.6319	10.6299
32	290	11.4173	11.4159	11.4193	11.4173
34	310	12.2047	12.2033	12.2067	12.2047
36	320	12.5984	12.5968	12.6006	12.5984
38	340	13.3858	13.3842	13.3880	13.3858
40	360	14.1732	14.1748	14.1754	14.1732

BEARING OD AND HOUSING DIMENSIONS—300 SERIES

Basic Bearing Number	Nominal Bearing OD	Bearing OD Specification		Housing Bore Specification	
		Max.	Min.	Max.	Min.
00	35	1.3780	1.3775	1.3790	1.3780
01	37	1.4567	1.4562	1.4577	1.4567
02	42	1.6535	1.6530	1.6545	1.6535
03	47	1.8504	1.8499	1.8514	1.8504
04	52	2.0472	2.0467	2.0484	2.0472
05	62	2.4409	2.4404	2.4421	2.4409
06	72	2.8346	2.8341	2.8358	2.8346
07	80	3.1496	3.1491	3.1508	3.1496
08	90	3.5433	3.5427	3.5447	3.5433
09	100	3.9370	3.9364	3.9384	3.9370
10	110	4.3307	4.3301	4.3321	4.3307
11	120	4.7244	4.7238	4.7258	4.7244
12	130	5.1181	5.1173	5.1197	5.1181
13	140	5.5118	5.5110	5.5134	5.5118
14	150	5.9055	5.9047	5.9071	5.9055
15	160	6.2992	6.2982	6.3008	6.2992
16	170	6.6929	6.6919	6.6945	6.6929
17	180	7.0866	7.0856	7.0882	7.0866
18	190	7.4807	7.4792	7.4821	7.4807
19	200	7.8740	7.8728	7.8758	7.8740
20	215	8.4646	8.4634	8.4664	8.4646
21	225	8.8583	8.8571	8.8601	8.8583
22	240	9.4488	9.4476	9.4506	9.4488
23	–	–	–	–	–
24	260	10.2362	10.2348	10.2375	10.2362
25	–	–	–	–	–
26	280	11.0236	11.0222	11.0249	11.0236
28	300	11.8110	11.8096	11.8123	11.8110
30	320	12.5984	12.5968	12.6006	12.5984
32	340	13.3858	13.3842	13.3880	13.3858
34	360	14.1732	14.1748	14.1754	14.1732
36	380	14.9606	14.9590	14.9628	14.9606
38	400	15.7480	15.7464	15.7802	15.7480
40	420	16.5354	16.5336	16.5379	16.5354

BEARING OD AND HOUSING DIMENSIONS—400 SERIES

Basic Bearing Number	Nominal Bearing OD	Bearing OD Specification		Housing Bore Specification	
		Max.	Min.	Max.	Min.
00	–	–	–	–	–
01	42	1.6535	1.6530	1.6545	1.6535
02	52	2.0472	2.0467	2.0484	2.0472
03	62	2.4409	2.4404	2.4421	2.4409
04	72	2.8346	2.8341	2.8358	2.8346
05	80	3.1496	3.1491	3.1508	3.1496
06	90	3.5433	3.5427	3.5447	3.5433
07	100	3.9370	3.9364	3.9384	3.9370
08	110	4.3307	4.3301	4.3321	4.3307
09	120	4.7244	4.7238	4.7258	4.7244
10	130	5.118	5.1173	5.1197	5.118
11	140	5.5118	5.5110	5.5134	5.5118
12	150	5.9055	5.9047	5.9071	5.9055
13	160	6.2992	6.2982	6.3008	6.2992
14	180	7.0866	7.0856	7.0882	7.0866
15	190	7.4803	7.4791	7.4821	7.4803
16	200	7.8740	7.8728	7.8758	7.8740
17	210	8.2677	8.2665	8.2695	8.2677
18	225	8.8583	8.8571	8.8601	8.8583
19	240	9.4488	9.4476	9.4506	9.4488
20	250	9.8425	9.8413	9.8443	9.8425
21	260	10.2362	10.2348	10.2375	10.2362
22	280	11.0236	11.0222	11.0249	11.0236
23	–	–	–	–	–
24	310	12.2047	12.2033	12.2067	12.2047
25	–	–	–	–	–
26	340	13.3858	13.3842	13.3880	13.3858
28	360	14.1732	14.1748	14.1754	14.1732
30	380	14.9606	14.9590	14.9628	14.9606
32	400	15.7480	15.7464	15.7402	15.7480
34	420	16.5354	16.5336	16.5379	16.5354
36	440	17.3228	17.3210	17.3253	17.3228
38	460	18.1102	18.1084	18.1127	18.1102
40	480	18.8976	18.8958	18.9001	18.8976

7

SHAFT ALIGNMENT

ROUGH SHAFT ALIGNMENT

Alignment of two pieces of equipment that are considered direct driven is critical to the service life of the equipment. Flexible coupling manufacturers brag about the amount of misalignment that their products can withstand. What is not mentioned is that bearings, mechanical seals, and other precision machine components will fail due to the forces produced by the misalignment. In other words, the machine will fail, but the coupling will still be okay!

The centerlines of the driver shaft and of the driven shaft of the machine need to be as close to the same as possible. The faster the RPM, the better the alignment that is needed.

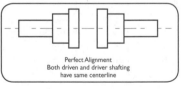

Perfect Alignment
Both driven and driver shafting
have same centerline

Misalignment occurs when the two shafts do not share the same centerline, because they are offset from each other or are at an angle to each other. One type of misalignment is commonly called *parallel* misalignment, whereby the shafts are parallel to each other but offset; the other is called *angular* misalignment, whereby the two shafts are at an angle to each other. In most cases, a combination of the two conditions exists.

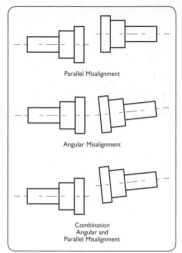

Parallel Misalignment

Angular Misalignment

Combination
Angular and
Parallel Misalignment

Alignment of 3600-RPM (or higher) machinery requires the use of dial indicators or a laser alignment tool to achieve the precision demanded by the speed. These methods can bring the shaft-to-shaft centerlines to be the same, within 0.001″ or less. Most manufacturers of alignment kits offer detailed instructions (or even classes) by which mechanics can learn these exacting procedures, but getting the machines into "rough" alignment is necessary before the precision alignment can be accomplished. It does not matter what type of precision alignment tooling is available, rough alignment is necessary to get the shafts "close enough" to allow the precision equipment to be brought into play. Furthermore, imagine a situation in which the higher-precision tools are not available and the machine needs to run to keep the process working or to allow the product to be manufactured for a key customer. In that case, rough alignment is better than nothing. Then the machine can be stopped at a future time so that precision (final) alignment can be accomplished. Rough alignment sets the stage for success at a later date.

Rough alignment begins with an investigation of the machine to determine whether there are any connected devices that would pull the machinery elements out of alignment in a process condition. Piping strain, conduit strain, a loose bedplate, or failure to provide for thermal expansion must be corrected; otherwise, the final alignment will be thrown off.

Investigation for a *soft foot condition* on the moveable element of the machine (usually the motor) and correction of this condition are required before any additional alignment corrections can be made. Soft foot occurs when the feet of the driver (or driven) shaft are not all on the exact same plane, thus permitting rocking to occur.

CORRECTING SOFT FOOT

Soft foot is a condition in which one of the feet of a machine (in many cases, the electric motor) does not sit flat on the baseplate. The foot or the base might have been warped. When a foot bolt is tightened, the machinery distorts. If not corrected, this condition can cause premature bearing problems, premature mechanical seal problems (on a pump), and continued difficulty in aligning the machine.

The two types of soft foot are *angular* and *parallel,* as shown here.

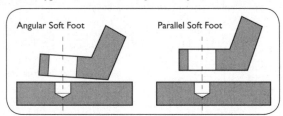

A quick way to check for the existence of soft foot is to first remove all the shims that might be found under the feet of the machine and then clean under the feet and on the baseplate to make sure no rust or debris is present. At that point, apply pressure at the four corners of the motor or pump (or other machinery element) and note any rocking. Rocking indicates that one of the four feet is not on the same level as the rest and thus soft foot is present.

There are two more-precise methods that can be used to detect a soft foot condition and correct for it. One uses a feeler gauge and the other uses a dial indicator.

Using a feeler gauge to detect soft foot

- Remove all the old shims and make sure that the foot and baseplate are clean and free of debris.
- Reinstall the drive unit; install the anchoring blots, but just slightly loosely at this point.
- Try to rock the unit to see whether soft foot is present. Rock across the diagonals.

- Use a feeler gauge under each foot. If at any point the feeler gauge passes under the foot and base, then soft foot is present.
- If a 0.002″ to 0.003″ feeler gauge is able to pass completely under the foot, a determination of the gap needs to be made so that the appropriate shim can be placed under the foot. This is the case for parallel soft foot.
- If a thicker feeler gauge can pass under the foot only to a certain point, then an angular soft foot condition has been identified.

Using dial indicators to detect soft foot

- Remove all the old shims and make sure that the foot and baseplate are clean and free of debris.
- Using the correct torque values, tighten the anchor bolts.
- Set the contact point of the dial indicator (zero the dial indicator) on one of the machinery feet.
- Slightly loosen the foot bolt and note the deflection on the dial indicator.
- Repeat steps for all the other mounting feet of the unit.
- Using the readings from the dial indicator, determine the soft foot.

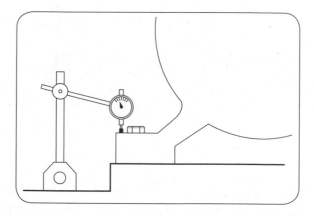

Correcting soft foot

Some of the possible methods of correcting the soft foot condition are as follows:

■ Remove the drive unit; either machine the baseplate to be a flat surface or machine the feet. Both of these methods are time consuming and costly.

■ Correct the issue through the alignment of the various components of the drive assembly.

■ If angular soft foot is the issue, the use of graduated, laminated, or stacking shims is another option.

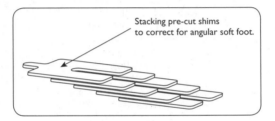

Stacking pre-cut shims to correct for angular soft foot.

■ In some cases, removing a portion of an existing precut shim facilitates the correction of an angular soft foot. Don't be afraid to use tin snips.

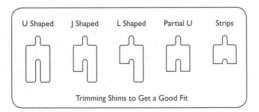

U Shaped   J Shaped   L Shaped   Partial U   Strips

Trimming Shims to Get a Good Fit

■ If the feet of the driver are badly worn or corroded, using a polymer product designed for alignment methods can assist in building the foot back up. Plastic steel and plastic aluminum are candidates for this. Welding and grinding are another alternative.

SHIM CHART

For the most part, when doing shaft alignment of a driver and driven machine, the driver is the component that will be shimmed. In most cases the driver is an electric motor. This chart shows the footprint of standard shims and indicates the proper shim for a given range of motor horsepower (HP). It is very important that the shim be wide enough to prevent rocking, which will make precision alignment impossible.

Standard Shim Sizing Chart						
Shim	A	B	C	D	G	H
L × W	2″ × 2″	3″ × 3″	4″ × 4″	6″ × 5″	7″ × 7″	8″ × 8″
Slot Width	5/8″	13/16″	1-1/4″	1-5/8 ×″	1-3/4″	2-1/4″
HP Range	1/4–15	10–60	50–200	150–1000	Over 1000	Over 1000

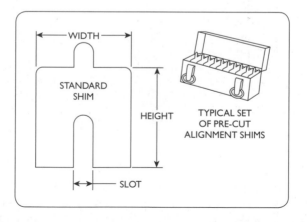

STRAIGHTEDGE AND FEELER GAGE ALIGNMENT

This method can be used to give a rough alignment in advance of using a dial indicator or laser to produce a precision alignment needed to complete the work.

It assumes that the face and outside diameters of the coupling halves are square and concentric with the coupling bores.

1. Check the angular misalignment using a micrometer or caliper. Measure from the outside of the flange to the opposite flange at four points 90 degrees apart. Do not rotate the coupling. Other methods of obtaining the same result make use of shims or a taper gauge to check the gap between both coupling halves. Measurement is again done at four points 90 degrees apart. With care, the multi-craftsman can bring the angular misalignment to below 1/64″ per inch of the coupling radius. With low-speed equipment (900 RPM or less), this may be adequate for a final alignment.

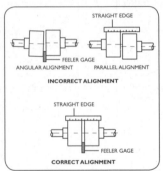

2. At four points 90 degrees apart (don't rotate the coupling), measure the parallel misalignment by laying a straightedge across one coupling half and measuring the gap between the straightedge and the opposite coupling half. Again, with care and using shims to check the distance between the straightedge and the coupling, a multi-craftsman can reduce the parallel misalignment to less than 0.005″.

8

V-BELT DRIVES

V-BELT DRIVES

The V-belt was developed in 1917 by John Gates of the Gates Rubber Company. It's called a V-belt because of its "V"-shaped cross section.

The V shape of the belt tracks in a mating groove in the sheave (or pulley), so that the belt cannot slip off. The belt also tends to wedge into the groove as the load increases—the greater the load, the greater the wedging action—improving torque transmission and making the V-belt a way of powering a machine. V-belts can be supplied at various fixed lengths or as a segmented section, whereby the segments are linked (spliced) to form a belt of the required length.

For high-power requirements, two or more V-belts can be joined side by side in an arrangement called a multi-V (or banded belt), running on matching multi-groove sheaves.

V-belts are identified by their cross sections. There are L series belts, classic belts (A through E* series), and narrow V (or wedge) belts.

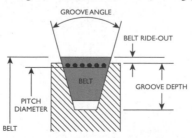

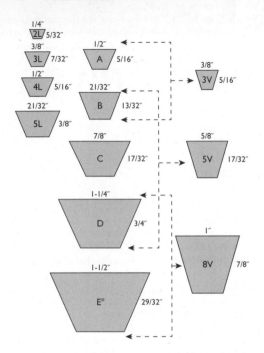

* The manufacturing of "E" Belts has been discontinued in 2011.

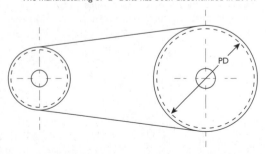

The pitch diameter (PD) of a sheave is not the outside diameter or the inside diameter. In fact, the pitch diameter is very difficult to measure directly. If you cut a belt and look at the end, you'll see a row of fibers near the outside surface. This is the tension-carrying part of the belt; the rest of the belt exists only to carry the forces from the sheave to and from

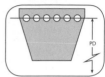

these fibers. The PD of any sheave is measured at these fibers. If you think about this for a moment, you'll see that the PD of a sheave depends not just on the sheave itself but on the width of the belt. If you put a B series belt on an A series sheave, it will ride higher than usual, increasing the effective PD.

V-BELT SHEAVE INSPECTION

It is important to understand that a V-belt transmits power from the driver to the driven sheave by the wedging action of the sidewalls of the belts against the sheave grooves. If the sheave grooves develop wear, the usual remedy (unfortunately) is increasing the belt tension to prevent the belts from squealing on start-up. This is a death sentence for the motor bearings and the machine driven by the belted system. For some reason, the fact that the sheave grooves look "bright and shiny" keeps uniformed multi-craftsmen from correcting the real problem—worn sheaves. Instead, new belts are installed and increased tension is applied.

Even in a clean environment, sheaves wear out in five years. The belts are made of rubber and fiber. Slowly they "sand" away the sheave sidewalls and remove the good geometry that needs to be present.

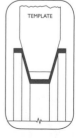

V-belt suppliers have sheave inspection tools available for performing sheave groove geometry checks to determine wear. A multi-craftsman should possess a set of these tools in his tool kit. Examination of belted systems for sheave wear is an essential activity in preventive maintenance. Replacing sheaves is a lot less expensive than replacing bearings!

SHEAVE SPEED RATIO CALCULATION

Most V-belt drives involve a speed change from driver to driven machine. Drives can be considered *step-up* (the driven turns faster than the driver) or *step-down* (the driven turns more slowly than the driver). The following schematic illustrates this type of drive.

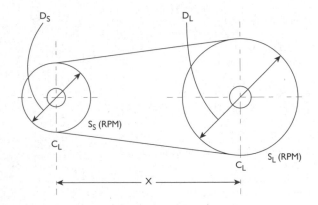

D_S = diameter of the small sheave
D_L = diameter of the large sheave
S_S = speed of the small sheave in RPM
S_L = speed of the large sheave in RPM
X = center-to-center distance between the sheaves

To find the size of a large sheave to produce a certain RPM when you know the size of the small sheave and the RPM of the small sheave:

$$D_L = \frac{D_S \times S_{S\,(RPM)}}{S_{L\,(RPM)}}$$

You have a step-down drive situation. The motor runs at 1750 RPM with a 6″-diameter sheave. You want the driven machine to operate at a speed of 1150 RPM. Calculate the sheave size for the driven machine as follows:

$$D_L = \frac{D_S \times S_{S\ (RPM)}}{S_{L\ (RPM)}}$$

$$D_L = \frac{6 \times 1750}{1150} = 9.130'' \text{ diameter}$$

A 9-1/4″-diameter sheave would probably be a stocked size. Variations of the formula include the following:

■ To find the size of a small sheave to produce a certain RPM when you know the size of the large sheave and the RPM of the large sheave:

$$D_S = \frac{D_L \times S_{L\ (RPM)}}{S_{S\ (RPM)}}$$

■ To find the RPM of the large sheave when you know the size of the small sheave, the RPM of the small sheave, and the size of the large sheave:

$$S_{L\ (RPM)} = \frac{D_S \times S_{S\ (RPM)}}{D_L}$$

■ To find the RPM of the small sheave when you know the size of the large sheave, the RPM of the large sheave, and the size of the small sheave:

$$S_{S\ (RPM)} = \frac{D_L \times S_{L\ (RPM)}}{D_S}$$

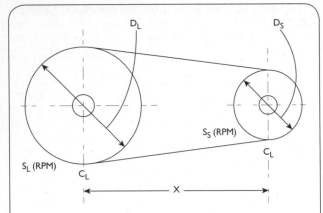

You have a step-up drive. The large sheave is attached to the motor, which runs at 1150 RPM. The large sheave is 12.5" in diameter and the small sheave (attached to the driven machine) is 7.375" in diameter. How fast will the driven machine turn?

$$S_{S\,(RPM)} = \frac{D_L \times S_{L\,(RPM)}}{D_S}$$

$$S_{S\,(RPM)} = \frac{12.5 \times 1150}{7.375} = 1949\,_{(RPM)}$$

What is the speed of the driven machine if a 3550-RPM motor is used as the driver?
Answer: 6017 RPM

V-BELT TENSIONING—FORCE DEFLECTION METHOD

To assist a multi-craftsman in correctly tensioning a V-belt, the use of a belt-tension measuring tool is recommended. Tension that is too low results in slippage and rapid wear of both belts and sheave grooves. Tension that is too high causes excessive stress and unnecessarily increases bearing loads. The following instructions explain the force deflection method of tensioning and give an example of proper use of a tension tester for correctly setting belt tension.

1. Measure the belt span (tangent to tangent) as shown here. (If this is too difficult to measure, the center-to-center distance between the two shafts can be used instead.)

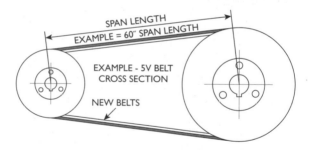

2. Set the *large* O-ring to the correct span length on the span scale of the tension tester. In this case, it is 60″.

3. Set the *small* O-ring at zero on the force scale of the tension tester.

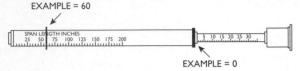

4. Place the metal end of the tester on one belt at the center of the belt span.

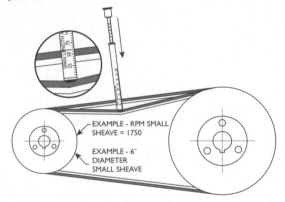

5. Apply force to the plunger until the bottom of the *large* O-ring is even with the top of any adjacent belt.

6. Read the force scale under the *small* O-ring to determine the force required to give the deflection.

7. Using the tables on the next two pages, compare the force scale reading with the correct value for either a new belt or an old belt.

8. If the force shown on the tester is less than the minimum, tighten the belts. If the force shown on the tester is greater than the maximum, loosen the belts.

Note: In the example given, the actual force for a new 5V-belt would be a minimum of 13.2 lb and a maximum of 19.8 lb. The actual reading of 14 lb would be acceptable.

| Belt Cross Section | Smaller Sheave Range | | Max./Min. Deflection Force | | | |
| | | | New Belt | | Old Belt | |
	Diameter	RPM	Max.	Min.	Max.	Min.
A	3 to 3-5/8	1000–2500	8.3	5.5	5.5	3.7
		2501–4000	6.3	4.2	4.2	2.8
	3-7/8 to 4-7/8	1000–2500	10.2	6.8	6.8	4.5
		2501–4000	8.6	5.7	5.7	3.8
	5 to 7	1000–2500	12.0	8.0	8.0	5.4
		2501–4000	10.5	7.0	7.0	4.7
B	3-3/8 to 4-1/4	850–2500	10.8	7.2	7.2	4.9
		2501–4000	9.3	6.2	6.2	4.2
	4-3/8 to 5-5/8	850–2500	11.0	7.9	7.9	5.3
		2501–4000	10.0	6.7	6.7	4.5
	5-7/8 to 8-5/8	850–2500	14.1	9.4	9.4	6.3
		2501–4000	13.4	8.9	8.9	6.0
C	7 to 9	500–1740	25.5	17.0	17.0	11.5
		1741–3000	20.7	13.8	13.8	9.4
	9-1/2 to 16	500–1740	31.5	21.0	21.0	14.1
		1741–3000	27.8	18.5	18.5	12.5
D	12 to 16	200–850	55.5	37.0	37.0	24.9
		851–1500	47.0	33.3	33.3	21.2
	18 to 20	200–850	67.8	45.2	45.2	30.4
		851–1500	57.0	38.0	38.0	25.6
E	21-5/8 to 24	100–450	71.0	47.0	47.0	31.3
		451–900	48.0	32.0	32.0	21.3

Quiz: A B series belt system has a small sheave of 7"-diameter operating at 1750 RPM. What are the minimum and maximum belt tensions for old belts?

Answer: ˙qๅ ㄣ˙6 = ɯnɯᴉxɐɯ 'qๅ ℰ˙9 = ɯnɯᴉuᴉW

Belt Cross Section	Smaller Sheave Range		Max./Min. Deflection Force			
			New Belt		Old Belt	
	Diameter	RPM	Max.	Min.	Max.	Min.
3V	2-1/4 to 2-3/8	1000–2500	7.4	4.9	4.9	3.3
		2501–4000	6.5	4.3	4.3	2.9
	2-5/8 to 3-5/8	1000–2500	7.7	5.1	5.1	3.6
		2501–4000	6.6	4.4	4.4	3.0
	4-18 to 6-7/8	1000–2500	11.0	7.3	7.3	4.9
		2501–4000	9.9	6.6	6.6	4.4
5V	4-3/8 to 6-5/8	500–1740	22.8	15.2	15.2	10.2
		1741–3000	19.8	13.2	13.2	8.8
	7-1/2 to 10-7/8	500–1740	28.4	18.9	18.9	12.7
		1741–3000	25.0	16.7	16.7	11.2
	11-7/8 to 16	500–1740	35.1	23.4	23.4	15.5
		1741–3000	32.7	21.8	21.8	14.6
8V	12-1/2 to 17	200–850	74.0	49.3	49.3	33.0
		851–1500	59.8	39.9	39.9	26.8
	18 to 22-3/8	200–850	88.8	59.2	59.2	39.6
		851–1500	79.0	52.7	52.7	35.3

Quiz: A 3V-belt system has a small sheave of 3"-diameter operating at 1150 RPM. What are the minimum and maximum belt tensions for new belts?

Answer: *Minimum = 5.1 lb, maximum = 7.7 lb.*

What if the speed is increased to 3550 RPM?

Minimum = 4.4 lb, maximum = 6.6 lb.

V-BELT TENSIONING—PERCENT ELONGATION METHOD

There are cases where V-belt tensioning of new belts is to be accomplished, but the necessary tools and information are not accessible to perform the force deflection method of tensioning. The force deflection method is preferred, but at two o'clock in the morning, with tools locked up in the storeroom, the big worry is to get a set of new belts tensioned somewhere in the right ballpark—not so tight that the bearings might fail due to overtensioning and not so loose that the belts slip and overheat. There is a procedure cited in older books for millwrights that allows this to be done. Keep three things in mind:

- ■ The procedure only involves the use of a tape measure and chalk (or a marker).
- ■ The procedure works *only* for the installation of new belts. Old belts have already been stretched and must be tensioned using a tension tester. The percent elongation method must not be used.
- ■ Since the procedure is a bit crude, it would be a good idea to "come back in the morning" and reset the tension using a tension tester, but the percent elongation method will work in a pinch.

First, install the belts and remove all the slack. Don't stretch the belts, but just snug them. Turn the sheaves to make sure.

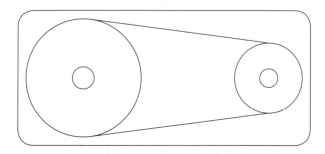

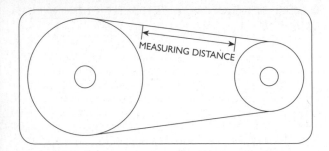

Now, using your chalk or marker, mark off a measured distance on the top of the belt that amounts to two-thirds of the distance available. It makes it easier if you choose a distance that ends in a zero, such as 20″, 60″, or 130″—the longer the distance, the better, to improve the accuracy.

Calculate a 2 percent increase in the measured distance and add it to the original distance. As an example, a 2 percent increase in a distance of 130″ is 2.6″ (130 × 0.02). Adding that to 130″ equals 132.6″. It's faster if you simply multiply by 1.02 to get the answer directly; 130″ × 1.02″ = 132.6″ gives you the same result—it's just faster.

Now begin to tension the belts and watch the measured distance between the two chalk marks change as the belt is stretched. When the new distance is equal to the measured distance plus 2 percent, the tension is set.

If 50″ was the original measured distance, then 2 percent of 50″ = 1″. Adding the two amounts gives 51″. Using the fast way: 50″ × 1.02 = 51″, and you are done.

Of course, there are a few rules governing this method. The following table provides some additional information:

Total stretch not less than 2 percent	
Normal Steady Loads	2% Stretch
High starting loads	Increase 1/2%
Extreme temperature or humidity	Increase 1/2%
Extreme dust or dirt	Increase 1/2%
Vertical shafts	Increase 1/2%
Belt speed over 6000 RPM	Increase 1/2%
Belt speed over 8000 RPM	Increase 1/2%
Total stretch not to exceed 5%	

You are installing a set of belts on a critical piece of plant equipment during the night shift. The belt-tension testing equipment is locked in the supervisor's desk, and she is on vacation. You decide to use the percent elongation method to tension the belts. You mark off a distance of 80" on the top of the belts, after removing all the slack. You notice that the area around the machine is very dusty and the drive is 150 HP with cross-the-line starting. What elongation would you use?

Answer: 2 percent plus 1/2 percent for the dusty area and another 1/2 percent for the high starting load (150 HP) = 3 percent total. 80" × 1.03 = 82.4" would be the new measured length to allow this percent elongation.

BELT LENGTH CALCULATION

The length of a V-belt for a particular drive system can also be calculated. You need to know the diameter of both sheaves as well as the center-to-center distance between the drive shafts.

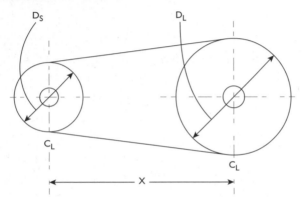

D_S = diameter of the small sheave
D_L = diameter of the large sheave
X = center-to-center distance between the sheaves
Π = 3.14
B_L = length of the V-belt

To find the length of belt for a given drive:

$$B_L = \frac{\pi \times (D_L + D_S)}{2} + 2\sqrt{X^2 + \left(\frac{D_L + D_S}{2}\right)^2}$$

Using a calculator makes it easy to do. We can work it in steps—it's just simple arithmetic.

If we have a belt system with a center-to-center distance of 75″ and the diameters of the sheaves are 12″ and 16″, it's matter of plugging the numbers into the formula to obtain the length of the V-belt needed to drive the machine. $X = 75''$; $D_S = 12''$; $D_L = 16''$

$$B_L = \frac{\pi \times (D_L + D_S)}{2} + 2\sqrt{X^2 + \left(\frac{D_L - D_S}{2}\right)^2}$$

$$B_L = \frac{\pi \times (16 + 12)}{2} + 2\sqrt{75^2 + \left(\frac{16 - 12}{2}\right)^2}$$

$$B_L = \frac{\pi \times (28)}{2} + 2\sqrt{5625 + \left(\frac{4}{2}\right)^2}$$

$$B_L = \frac{3.14 \times (28)}{2} + 2\sqrt{5625 + (2)^2}$$

$$B_L = \frac{87.92}{2} + 2\sqrt{5629}$$

The square root of 5629 is equal to approximately 75.

$$B_L = 43.96 + (2 \times 75)$$
$$B_L = 43.96 + 150 = 193.96''$$

You would choose the belt that came closest to a stock length of 194''—most probably a 195'' belt would be the closest length available from most belt companies.

STANDARD V-BELT LENGTHS

Belt #	A		B		C		D		E	
Including Nominal Length	Pitch Length	Outside Length	Pitch Length	Outside Length	Pitch Length	Outside Length	Pitch Length	Outside Length	Pitch Length	Outside Length
26	27.3	28.0	—	—	—	—	—	—	—	—
31	32.3	33.0	—	—	—	—	—	—	—	—
33	34.3	35.0	—	—	—	—	—	—	—	—
35	36.3	37.0	36.8	38.0	—	—	—	—	—	—
38	39.3	40.0	39.8	41.0	—	—	—	—	—	—
42	43.3	44.0	43.8	45.0	—	—	—	—	—	—
46	47.3	48.0	47.8	49.0	—	—	—	—	—	—
48	49.3	50.0	49.8	51.0	—	—	—	—	—	—
51	52.3	53.0	52.8	54.0	53.9	89.0	—	—	—	—
53	54.3	55.0	—	—	—	—	—	—	—	—
55	56.3	57.0	56.8	58.0	—	—	—	—	—	—
60	61.3	62.0	61.8	63.0	62.9	64.0	—	—	—	—
62	63.3	64.0	63.8	65.0	—	—	—	—	—	—
64	65.3	66.0	65.8	67.0	—	—	—	—	—	—
66	67.3	68.0	67.8	69.0	—	—	—	—	—	—
68	69.3	70.0	69.8	71.0	70.9	72.0	—	—	—	—
71	72.3	73.0	72.8	74.0	—	—	—	—	—	—
75	76.3	77.0	76.8	78.0	77.9	79.0	—	—	—	—
78	79.3	78.0	79.8	81.0	—	—	—	—	—	—
80	81.3	82.0	—	—	—	—	—	—	—	—
81	—	—	82.8	84.0	83.9	85.0	—	—	—	—
83	—	—	84.8	86.0	—	—	—	—	—	—
85	86.3	87.0	86.8	88.0	87.9	89.0	—	—	—	—
90	91.3	92.0	91.8	93.0	92.9	94.0	—	—	—	—
96	97.3	98.0	—	—	98.9	100.0	—	—	—	—
97	—	99.0	98.8	100.0	—	—	—	—	—	—
105	106.3	107.0	106.8	108.0	107.9	109.0	—	—	—	—
112	113.3	114.0	113.8	115.0	114.9	116.0	—	—	—	—
120	121.3	122.0	121.8	123.0	122.9	124.0	123.3	125.0	—	—

Belt #	A		B		C		D		E	
Including Nominal Length	Pitch Length	Outside Length	Pitch Length	Outside Length	Pitch Length	Outside Length	Pitch Length	Outside Length	Pitch Length	Outside Length
128	129.3	130.0	129.8	131	130.9	132.0	131.3	133.0	—	—
136	—	—	137.8	139.0	138.9	140.0	—	—	—	—
144	—	—	145.8	147.0	146.9	148.0	147.3	149.0	—	—
158	—	—	159.8	161.0	160.9	162.0	161.3	163.0	—	—
162	—	—	—	—	164.9	166.0	165.3	167.0	—	—
173	—	—	174.8	176.0	175.9	177.0	176.3	178.0	—	—
180	—	—	181.8	183.0	182.9	184.0	183.3	185.0	184.5	187.5
195	—	—	196.8	198.0	197.9	199.0	198.3	200.0	199.5	202.5
210	—	—	211.8	213.0	212.9	214.0	213.3	215.0	214.5	217.5
240	—	—	240.3	241.5	240.9	242.0	240.8	242.0	241.0	244.0
270	—	—	270.3	271.5	270.9	272.0	270.8	272.5	271.0	274.0
300	—	—	300.3	301.5	300.9	302.0	300.8	302.5	301.0	304.0
330	—	—	—	—	—	—	330.8	332.5	331.0	334.0
360	—	—	—	—	360.9	362.0	360.8	362.5	361.0	364.0
390	—	—	—	—	390.9	392.0	390.8	392.5	391.0	394.0
420	—	—	—	—	420.9	422.0	420.8	422.5	421.0	424.0
480	—	—	—	—	—	—	480.8	482.5	481.0	484.0
540	—	—	—	—	—	—	540.8	542.5	541.0	544.0
600	—	—	—	—	—	—	600.8	602.5	601.0	604.0
660	—	—	—	—	—	—	—	—	661.0	664.0

V-Belt Facts: The preferred center-to-center distance is larger than the largest pulley diameter but less than three times the sum of both pulleys. The optimal speed range is 1000 to 7000 ft/min. Fatigue is the culprit for most belt problems. This wear is caused by stress from rolling around the pulleys. High belt tension, excessive slippage, and belt overloads caused by shock, vibration, or belt slapping all contribute to belt fatigue.

3V	
Belt #	Pitch Length
250	25.0
280	28.0
265	26.5
300	30.0
315	31.5
335	33.5
355	35.5
375	37.5
400	40.0
425	42.5
450	45.0
475	47.5
500	50.0
530	53.0
560	56.0
600	60.0
630	63.0
670	67.0
710	71.0
750	75.0
800	80.0
850	85.0
900	90.0
950	95.0
1000	100.0
1060	106.0
1120	112.0
1180	118.0
1250	128.0
1320	132.0
1400	140.0

5V	
Belt #	Pitch Length
500	50.0
530	53.0
560	56.0
600	60.0
630	63.0
670	67.0
710	71.0
750	75.0
800	80.0
850	85.0
900	90.0
950	95.0
1000	100.0
1060	106.0
1120	112.0
1180	118.0
1250	128.0
1320	132.0
1400	140.0
1500	150.0
1600	160.0
1700	170.0
1800	180.0
1900	190.0
2000	200.0
2120	212.0
2240	224.0
2360	236.0
2500	250.0
2650	265.0
2800	280.0
3000	300.0
3150	315.0
3350	335.0
3550	355.0

8V	
Belt #	Pitch Length
1000	100.0
1060	106.0
1120	112.0
1180	118.0
1250	128.0
1320	132.0
1400	140.0
1500	150.0
1600	160.0
1700	170.0
1800	180.0
1900	190.0
2000	200.0
2120	212.0
2240	224.0
2360	236.0
2500	250.0
2650	265.0
2800	280.0
3000	300.0
3150	315.0
3350	335.0
3550	355.0
3750	375.0
4000	400.0
4250	425.0
4500	450.0
5000	500.0

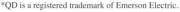

9

MECHANICAL BUSHINGS

MECHANICAL BUSHINGS

There are two common types of bushings used to attach a V-belt sheave (or chain sprocket) to a shaft: the QD® bushing* and the Taper Lock® bushing.†

The main difference between the two is that the QD-style bushing has a flange around the outside diameter, while the taper lock bushing has straight sides on the OD. Many people refer to both types as taper locks because both use the tapered wedging action to lock to the shaft. The taper lock bushing, with its straight sides, uses an internal hex-head cap screw to drive the bushing into the bore of the component being installed (sheave, sprocket, etc.). Be careful when installing these screws. The screws go into the blind holes in the bushing that are threaded in the installed component. The holes with threads on the bushing are for removal only. Also be aware that the appearance of a flange on the outside of the bushing doesn't necessarily mean it's a QD style. The Browning Split Taper bushing also has a flange, and the two are not interchangeable. The Browning style can be identified by a key on both the inside and the outside of the bushing, and the flange is solid. The QD style has a split that continues through the flange.

QD® BUSHING

*QD is a registered trademark of Emerson Electric.

†Taper Lock is a registered trademark of Reliance Electric.

MECHANICAL BUSHINGS—QD INSTALLATION

Installation of QD Bushings:

1. Clean all parts of the bushing and bore of the hub, removing any oil, lacquer, or dirt. *Do not use antiseize or any other type of lubricant on tapered cone surfaces—the use of lubricants can cause hub fracture.* File away any burrs.

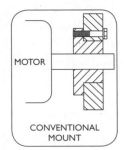

2. For a conventional mounting, assemble the sheave (or sprocket) and bushing combination by sliding the sheave (sprocket) tapered bore into position over the mating tapered bushing surface. Align the unthreaded holes in the sheave (or sprocket) hub with the threaded holes in the flange of the bushing.

With the lock washers installed, tighten the cap screws by hand. The sheave (or sprocket) and bushing assembly mount onto the shaft, with the bushing flange facing inward. Some assemblies allow a reverse mount procedure. In this configuration, the bushing flange faces outward but still allows the cap screw installation from the outside of the assembly. The cap screws fit through the unthreaded holes of the bushing flange and into the threaded holes of the sheave (or sprocket) hub.

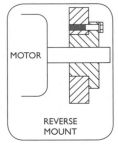

3. With the key resting in the shaft keyway, position the assembly onto the shaft, allowing for a small axial movement of the sheave (or sprocket), which will occur during the tightening process.

When installing large or heavy parts in a conventional mount, it is easier to mount the key and bushing onto the shaft first, then place the sheave (sprocket) on the bushing and align the holes.

4. Alternately tighten the cap screws until the sheave (or sprocket) and the bushing tapers are completely seated. Use approximately half of the bolt torque recommended in the following table.

QD Bushing Bolt Torque Chart

Bushing Style	Bolts (inches)		Torque Wrench ft-lb
	Quantity	Size	
H	2	1/4-20 × 3/4	7.9
JA	3	10-24 × 1	4.5
SH, SDS	3	1/4-20 × 1-3/8	9.0
SD	3	1/4-20 × 1-7/8	9.0
SK	3	5/16-18 × 2	15.0
SF	3	3/8-16 × 2	30.0
E	3	1/2-13 × 2-3/4	60.0
F	3	9/16-12 × 3-5/8	75.0
J	3	5/8-11 × 4-1/2	135.0
M	4	3/4-10 × 6-3/4	225.0
N	4	7/8-9 × 8	300.0
P	4	1-8 × 9-1/2	450.0
W	4	1-1/8—7 × 11-1/2	600.0
S	5	1-1/4—7 × 15-1/2	750.0

5. Check the alignment and axial sheave run-out (wobble) and correct.
6. Continue to alternately tighten cap screws to the torque values shown in the table.
7. Tighten the setscrew, if available, to hold the key.

Removal of QD Bushings:

1. Loosen and remove all mounting bolts. Do not forget the setscrew for the key, if so equipped.

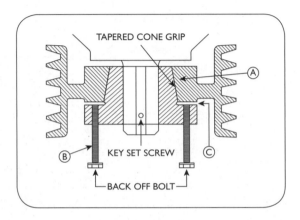

2. Insert cap screws (B) into all threaded jackscrew holes.

3. Loosen the bushing by first tightening the screw farthest from the bushing saw slot, then alternately tighten the remaining screws. Keep tightening in small but equal increments until the tapered sheave (A) disengages from the hub (C).

4. Use care to keep pressure even during the disassembly procedure, since unequal pressure can break the bushing flange, making further removal impossible without destroying the sheave (or sprocket).

5. Inspect the components carefully, looking for cracks, wear, or rough or dinged surfaces.

MECHANICAL BUSHINGS—TAPER LOCK INSTALLATION

Installation of Taper Lock Bushings:

1. Clean all parts of the bushing and bore of the hub, removing any oil, lacquer, or dirt. *Do not use antiseize or any other type of lubricant on tapered cone surfaces—the use of lubricants can cause hub fracture.* Install the bushing in the hub and match half holes to make complete holes (each complete hole will be threaded on one side only).

2. Oil the thread and either the ends of the setscrews or under the heads of the cap screws. Install the screws loosely in the holes that are threaded on the hub.

TAPER LOCK® BUSHING

3. Make sure that the bushing is free in the hub. Slip the assembly onto the shaft and align in the desired position.

4. Tighten the screws evenly and alternately until the part has tightened.

5. Using a block or sleeve, hammer the large end of the bushing. Retighten the screws using the correct torque. Repeat this procedure until the screws no longer turn. Fill the remaining holes with grease or modeling clay to prevent dirt buildup.

To Remove:

1. Remove all screws. Oil the thread and either the ends of the setscrews or under the heads of the capscrews.

2. Insert the screws into the hole(s) that are threaded on the bushing side. Note that there will be one screw left over.

3. Tighten the screws alternately until the bushing is loose in the hub. It may be necessary to tap on the hub to loosen the bushing.

10

ELECTRICITY

THE LAWS OF ELECTRICITY

The German physicist Georg Ohm, published a paper in 1827 that related the voltage, current, and resistance of electrical circuits; his thesis is often called Ohm's law. Ohm's law states that the current through a conductor between two points is directly proportional to the potential difference or voltage across the two points and is inversely proportional to the resistance between them. This law is the basis of many electrical calculations. Expressed mathematically:

$$E = I \times R$$

I is the current flowing through the conductor, measured in amperes; E is the potential difference across the conductor, measured in volts; and R is the resistance of the conductor, measured in ohms.

The chart shown here contains commonly used formulas related to Ohm's law. It allows users to quickly find the correct formulas to solve for volts, amps, ohms, or watts (power) if other components of the circuit are known.

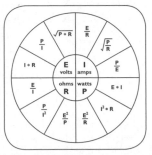

Some examples demonstrate the usefulness of the chart.

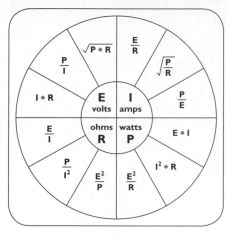

If a kitchen toaster oven is rated on the nameplate at 1500 watts if used on a 110-volt circuit, how much current will be drawn by this appliance? Choose the correct formula from the chart by looking in the sector where the unknown quantity is located. In this case the unknown quantity is the current draw, or amperage.

The known quantities are the watts (P) and the voltage (E). The correct formula is:

$$I = \frac{P}{E} = \frac{1500}{110} = 13.6 \text{ amps}$$

What is the resistance of the toaster oven, measured in ohms?

$$R = \frac{E}{I} = \frac{110}{13.6} = 8.08 \text{ ohms}$$

Electrical shocks can be dangerous. There is an old, but true, saying: "It's not the voltage that kills, it's the current." Here is a chart showing values of current that might pass through a human being and the effects.

The first real indication of electrical shock takes place at 0.001 amps, at which point a tiny bit of electricity can be felt. Above 0.01 amp the muscles of the body lock and a person cannot let go of the wire or device that is delivering the shock. As the current value increases, breathing becomes labored and finally stops.

As the current approaches 0.1 amp, ventricular fibrillation of the heart—an unco-ordinated twitching of the walls of the ven-tricles—occurs, which results in death. Oddly enough, when cur-rent values are above 0.2 amp the muscular contractions are so severe that the heart is forcibly clamped during the shock. This

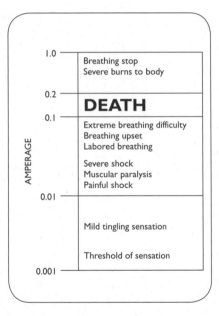

clamping protects the heart from going into ventricular fibrillation, and the victim's chances for survival are good if artificial respiration tech-niques are employed quickly, as soon as the victim is removed from the source of the shocking current. Some level of burn treatment is required in such cases.

The skin of a relatively dry human being has a resistance of about 100,000 ohms. If the person touches the terminals of a 9-volt battery, how much current will flow?

$$I = \frac{E}{R} = \frac{9}{100,000} = 0.00009 \text{ amp}$$

Not much chance of a mishap here.

What danger is present if the same person touches a bare wire in a 120-volt circuit?

$$I = \frac{E}{R} = \frac{120}{100,000} = 0.0012 \text{ amp}$$

The person would most likely feel a tingle of electricity.

The resistance of the skin of a person who had just stepped out of a shower and was soaking wet is about 100 ohms. What would be the result if this person touched the 9-volt battery and the 120-volt circuit?

$$I = \frac{E}{R} = \frac{9}{100} = 0.09 \text{ amp}$$

$$I = \frac{E}{R} = \frac{12}{100} = 1.2 \text{ amp}$$

The effects would be fairly damaging, even from a device as small as the simple 9-volt battery found in a smoke alarm. Water lowers skin resistance by 1000 times. The upshot: Around electricity, stay dry!

11

ELECTRIC MOTORS

UNDERSTANDING AND TROUBLESHOOTING INDUSTRIAL INDUCTION MOTORS

The most common form of industrial induction motor is the squirrel-cage motor. The name was derived because the rotor resembles the wheel of a squirrel cage. Its universal use lies in its mechanical simplicity, its ruggedness, and the fact that it can be manufactured with characteristics to suit most industrial requirements.

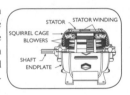

This motor consists of two basic components: the *rotor,* which spins, and the *stator,* which remains stationary. The rotor is held in proper position by bearings that are fitted into endplates (often called *end bells*). Between the rotor and the stator is a small air gap that allows clearance for rotation to take place. Years ago, motors were produced with a very large air gap. When electricity was not so expensive the inefficiency of the motor, caused by this large gap, was not important, but as electricity prices and metal prices increased, more efficient motors were designed with smaller and smaller air gaps. The most common designs of squirrel-cage motors are the *U-frame* and the *T-frame.* The National Electrical Manufacturers Association (NEMA) publishes standards covering dimensions for shaft size, height, and bolt-down holes for specific frame numbers. Motors made by different manufacturers must comply with this set of specifications. Hence, motors from different manufacturers can be interchanged if the frame numbers are the same.

Symptom	Possible Cause
Motor will not start.	Overload has tripped. Power is not connected. Fuses are faulty (open). Voltage is low. Terminal lead connections are loose. Driven machine is locked up. There is an open circuit in the stator winding. There is a short circuit in the stator winding. Bearings are stiff. Grease is too stiff.
Motor is noisy.	Motor is running at single phase. Electrical load is unbalanced. Shaft is bumping (on sleeve bearing motors). Vibration carries over from driven machine. Air gap is not uniform. Anti-friction bearings are noisy. Shaft has loose parts or loose rotor. Rotor is rubbing on stator. Object is caught between fan and end bell. Coupling is loose. Motor is loose on foundation.
Motor is at higher-than-normal temperature or is smoking.	There is an overload. Electrical load is unbalanced (fuse blown, faulty control). Ventilation is restricted. Voltage and frequency are incorrect. Motor is stalled by driven machine or tight bearings. Stator windings have shorted. Stator windings are grounded. Rotor winding has loose connections. Belts are too tight. Incorrect motor was used for rapid reversing service.

Symptom	Possible Cause
Motor bearings (antifriction bearings) are hot.	Wrong grade of grease was used. Insufficient amount of grease was used. Too much grease was used, causing churning. Grease is contaminated by foreign material. Bearings are misaligned. Bearings are damaged (corrosion, spalling, etc.). End shields are loose or cocked. Belt tension or thrust is excessive. Shaft is bent.
Motor bearings (sleeve bearings) are hot.	Insufficient oil was used. Oil is contaminated by foreign material, or a poor grade of oil was used. Oil rings are rotating slowly or not rotating at all. Motor is tilted too far. Oil rings are bent or otherwise damaged in housing. Oil rings are out of slot. Motor is tilted, causing end thrust. Bearings are defective, or shaft is rough.
Motor overload keeps tripping.	Load is too much for motor. Ambient temperature is too high. Thermal overload is bad—replace with new one. Winding has shorted or grounded.
Motor runs in wrong direction.	"T" leads to motor are reversed; change any two.
Motor runs but then dies down.	Voltage dropped. Load increased.

IMPROVING MOTOR BEARING LIFE

Most industrial equipment uses an electric motor as a driver. It's a little-known fact that electric motors actually present a relatively easy duty for shaft bearings. The motor rotor is lightweight, yet, because of its large shaft diameter, the bearings are large. For example, the bearings supporting the 140-lb rotor for 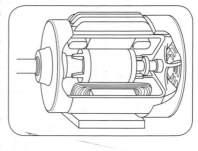 a typical 40-HP 1750-RPM industrial motor are so large that they have an L-10 minimum design fatigue life of 3000 years. Translated into shop English, this means that only 10 percent of the bearings are statistically expected to fail from fatigue after *3000 years* of operation and the rest (90 percent) will last even longer!

Actual operating experience, however, does not support such optimistic estimates of motor bearing life. In real industrial environments, bearing failure is rarely caused by fatigue; it is caused by less-than-ideal lubrication. Because of contaminated lubrication, bearings fail well before they serve their theoretical fatigue life. There are many reasons for less-than-ideal bearing lubrication. Lubricants can leak out; chemical attacks or thermal conditions can decompose or break down lubricants; lubricants can become contaminated with nonlubricants such as water, dust, or rust from the bearings themselves. Fortunately, these lubrication problems can be eliminated. Motor bearings can last virtually forever by simply providing an ideal contamination-free, well-lubricated bearing environment. Most motors, from the factory, provide lip seals to keep lubrication in and water out or sealed-for-life bearings prepacked with grease.

It is a fact that lip seals invariably wear out well before the bearing fails, and sealed bearings inherently foreshorten the life of a bearing to the service life of the contained grease (usually only about 3000 to

5000 hours for most industrial services). Here are some proven tips to get much longer service life out of motor bearings and reduce downtime and costs at your facility.

■ *Lubricate bearings at correct intervals.*

Overlubrication ruins motor bearings. Too much grease causes overheating of the bearings. The lubrication instructions supplied by the motor manufacturer specify the quantity and frequency of lubrication. As a rule of thumb, two-pole motors (3600 RPM) should be greased twice a year; four-pole and slower motors (1800, 1200, 900, 600 RPM), only once a year.

■ *Change to use the best available grease.*

The most commonly used bearing grease is polyurea-based, a low-cost, low-performance, highly compatible lubricant. However, it does not handle water well—a serious drawback. It reacts readily with water and loses its ability to lubricate bearings. Consider using a synthetic-based aluminum complex grease. A high-quality grease pays for its additional cost in reduced motor downtime and repair costs.

■ *Make changes to keep out moisture and dirt.*

Here's the problem: When the motor runs, unless it is being hosed down or operates in a humid environment, reasonably shielded motor bearings don't become contaminated with moisture. But when the *motor is shut down,* moisture and condensation can collect on the surface of the bearing components. Eventually, this water breaks through the oil and grease barrier, contacts the metal parts of the bearing, and produces tiny particles of iron oxide. These rust particles mix with the grease to form a grinding compound. It's like adding sand.

Close shaft–to–end bell clearances cannot stop the movement of humid air. Contact seals lose their contact, resulting in large gaps that allow movement of air and water vapor across the bearing.

Vapor-blocking bearing isolators, such as the one illustrated, are among the more successful devices presently available to prevent water vapor from entering a stationary bearing. When the motor shaft is rotating, the isolator opens, eliminating the possibility of friction and wear. However, when the shaft stops, the isolator closes, preventing movement of air, water, or dirt across its face—a perfect solution to help motor bearings last 5 to 10 times as long as those with typical original equipment manufacturer (OEM) sealing components. Since these types of isolators do not wear from rotating friction, the seal may last indefinitely.

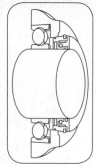

Although not intended as such, a bearing isolator could serve as an emergency sleeve bearing if the primary bearing fails, possibly preventing damage to the motor's stator and rotor. In emergency situations, the bearing isolator can allow continued operation for a short time and still prevent the need to rewind the motor when the bearing is replaced.

ELECTRIC MOTOR INSULATION VALUES

If a motor is chosen properly for a specific application, the insulation should prove serviceable for periods of upward of 20 years. The following table shows the various classes of insulation available as well as the hot-spot temperature. Overheating of the motor beyond these values causes rapid deterioration of the insulation, resulting in a burned-up motor that exhibits a pungent, "burned insulation" odor.

Motor Insulation Classes			
Class	Hot-Spot Temperature		Notes
	°C	°F	
A	105°C	221°F	Rarely used anymore
B	130°C	266°F	Hydroscopic
F	155°C	311°F	Non-hydroscopic
H	180°C	356°F	Non-hydroscopic

Insulation Class B is the most common off-the-shelf class used in motors. This type of insulation is intended to be used in dry locations. Class B insulation is hydroscopic, meaning that it absorbs water, which can quickly degrade the windings. Motors with Class B insulation are unsuited for wet environments such as outside service, rooftops, or locations near a wet process. A better selection for such applications is Class F or H. Often, motors purchased in an emergency are Class B, since vendors generally have them on the shelf, and these will fail quickly if the original motor was Class F or H. It's worth looking at the motor nameplate and noting the insulation class specified before purchasing a replacement motor.

> *TIP: If a Class B motor is sent out for rewinding, it is a fairly cheap option to have it rewound with Class F. It's an easy way to improve a motor and make it more versatile*

INDUSTRIAL MOTOR DIMENSION CHARTS

A key part of motor interchangeability has been the standardization of frame sizes. This means that the same horsepower, speed, and enclosure will normally have the same frame size from different motor manufacturers. Thus, a motor from one manufacturer can be replaced with a similar motor from another company, provided they are both in standard frame sizes.

Fractional Horsepower Motors

The term *fractional horsepower* is used to cover frame sizes that have two-digit designations. The frame sizes normally associated with fractional-horsepower motors are 42, 48, and 56. Each frame size designates a particular shaft height, shaft diameter, and face- or base-mounting hole pattern. Specific frame assignments are not made 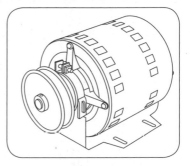 by horsepower and speed, so it is possible that a particular horsepower and speed combination might be found in three different frame sizes. When replacement is required, it is essential that the frame size be known as well as the horsepower, speed, and enclosure. The two-digit frame number is based on the shaft height in sixteenths of an inch. A 48 frame motor has a shaft height of 48″ divided by 16″ or 3″. Similarly, a 56 frame motor has a shaft height of 3-1/2″. The largest of the current fractional horsepower frame sizes is a 56 frame, which is available in horsepowers greater than fractionals. The 56 frame motors are built up to 3 HP and, in some cases, 5 HP. For this reason, calling motors with two-digit frame sizes "fractionals" can be misleading.

Integral Horsepower Motors

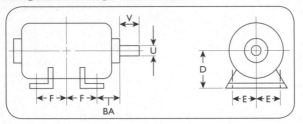

The term *integral horsepower* refers to those motors that have three-digit frames, such as 143T or larger. When dealing with these frame sizes one rule of thumb is handy: The centerline shaft height ("D" dimension) above the bottom of the base is the first two digits of the frame size divided by 4. For example, a 254T frame would have a shaft height of $25 \div 4 = 6.25″$. Although the last digit does not directly relate to a dimension in inches, larger numbers do indicate that the rear bolt holes are moved farther away from the shaft end bolt holes (the "F" dimension becomes larger).

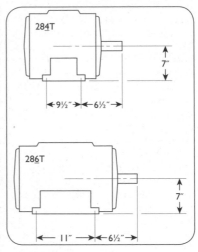

The use of the letter "S" in a motor frame designation indicates that the motor has a *short shaft*. Short shaft motors have shaft dimensions that are smaller than the shafts associated with the normal frame size. Short shaft motors are designed to be directly coupled to a load through a flexible coupling. They are not supposed to be used for applications where belts are used to drive the load.

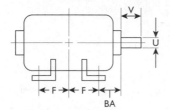
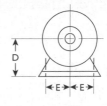

Frame	U	D	E	F	V	BA
42	3/8	2-5/8	1-3/4	2-7/32	1-1/8	2-1/16
48	1/2	3	2-1/8	1-3/8	1-1/2	2-1/2
48H	5/8	3	2-1/8	2-3/8	1-1/2	2-1/2
56	5/8	3-1/2	2-7/16	1-1/2	1-7/8	2-3/4
56H	5/8	3-1/2	2-7/16	2-1/2	1-7/8	2-3/4
66	3/4	4-1/8	2-15/16	2-1/2	2-1/4	3-1/8
143T	7/8	3-1/2	2-3/4	2	2-1/4	2-1/4
145T	7/8	3-1/2	2-3/4	2-1/2	2-1/4	2-1/4
182	7/8	4-1/2	3-3/4	2-1/4	2	2-3/4
182T	1-1/8	4-1/2	3-3/4	2-1/4	2-1/2	2-3/4
184	7/8	4-1/2	3-3/4	2-3/4	2	2-3/4
184T	1-1/8	4-1/2	3-3/4	2-3/4	2-1/2	2-3/4
203	3/4	5	4	2-3/4	2	3-1/8
204	3/4	5	4	3-1/4	2	3-1/8
213	1-1/8	5-1/4	4-1/4	2-3/4	2-3/4	3-1/2
213T	1-3/8	5-1/4	4-1/4	2-3/4	3-1/8	3-1/2
215	1-1/8	5-1/4	4-1/4	3-1/2	2-3/4	3-1/2
215T	1-3/8	5-1/4	4-1/4	3-1/2	3-1/8	3-1/2
224	1	5-1/4	4-1/2	3-3/8	2-3/4	3-1/2
225	1	5-1/2	4-1/2	3-3/4	2-3/4	3-1/2
254	1-1/8	6-1/4	5	4-1/8	3-1/8	4-1/4
254U	1-3/8	6-1/4	5	4-1/8	3-1/2	4-1/4
254T	1-5/8	6-1/4	5	4-1/8	3-3/4	4-1/4
256U	1-3/8	6-1/4	5	5	3-1/2	4-1/4
256T	1-5/8	6-1/4	5	5	3-3/4	4-1/4
284	1-1/4	7	5-1/2	4-3/4	3-1/2	4-3/4
284U	1-5/8	7	5-1/2	4-3/4	4-5/8	4-3/4
284T	1-7/8	7	5-1/2	4-3/4	4-3/8	4-3/4
284TS	1-5/8	7	5-1/2	4-3/4	3	4-3/4
286U	1-5/8	7	5-1/2	5-1/2	4-5/8	4-3/4
286T	1-7/8	7	5-1/2	5-1/2	4-3/8	4-3/4
286TS	1-5/8	7	5-1/2	5-1/2	3	4-3/4

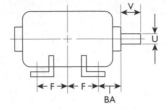
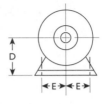

Frame	U	D	E	F	V	BA
324T	2-1/8	8	6-1/4	5-1/4	5	5-1/4
324	1-5/8	8	6-1/4	5-1/4	4-5/8	5-1/4
324U	1-7/8	8	6-1/4	5-1/4	5-3/8	5-1/4
324S	1-5/8	8	6-1/4	5-1/4	3	5-1/4
326	1-5/8	8	6-1/4	6	4-5/8	5-1/4
326S	1-5/8	8	6-1/4	6	3	5-1/4
326U	1-7/8	8	6-1/4	6	5-3/8	5-1/4
326T	2-1/8	8	6-1/4	6	5	5-1/4
326TS	1-7/8	8	6-1/4	6	3-1/2	5-1/4
364	1-7/8	9	7	5-5/8	5-3/8	5-7/8
364S	1-5/8	9	7	5-5/8	3-1/2	5-7/8
364U	2-1/8	9	7	5-5/8	6-1/8	5-7/8
364US	1-7/8	9	7	5-5/8	3-1/2	5-7/8
364T	2-3/8	9	7	5-5/8	5-5/8	5-7/8
364TS	1-7/8	9	7	5-5/8	3-1/2	5-7/8
365	1-7/8	9	7	6-1/8	5-3/8	5-7/8
365S	1-5/8	9	7	6-1/8	3-1/2	5-7/8
365U	2-1/8	9	7	6-1/8	6-1/8	5-7/8
365US	1-7/8	9	7	6-1/8	3-1/2	5-7/8
365T	2-5/8	9	7	6-1/8	5-5/8	5-7/8
365TS	1-7/8	9	7	6-1/8	3-1/2	5-7/8
404	2-1/8	10	8	6-1/8	6-1/8	6-5/8
404S	1-7/8	10	8	6-1/8	3-1/2	6-5/8
404U	2-3/8	10	8	6-1/8	6-7/8	6-5/8
404US	2-1/8	10	8	6-1/8	4	6-5/8
404T	2-7/8	10	8	6-1/8	7	6-5/8
404TS	2-1/8	10	8	6-1/8	4	6-5/8
405	2-1/8	10	8	6-7/8	6-1/8	6-5/8
405S	1-7/8	10	8	6-7/8	3-1/2	6-5/8
405U	2-3/8	10	8	6-7/8	6-7/8	6-5/8
405US	2-1/8	10	8	6-7/8	4	6-5/8
405T	2-7/8	10	8	6-7/8	7	6-5/8
405TS	2-1/8	10	8	6-7/8	4	6-5/8

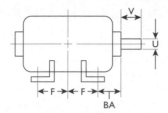
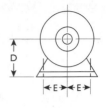

Frame	U	D	E	F	V	BA
444	2-3/8	11	9	7-1/4	6-7/8	7-1/2
444S	2-1/8	11	9	7-1/4	4	7-1/2
444U	2-7/8	11	9	7-1/4	8-3/8	7-1/2
444US	2-1/8	11	9	7-1/4	4	7-1/2
444T	3-3/8	11	9	7-1/4	8-1/4	7-1/2
444TS	2-3/8	11	9	7-1/4	4-1/2	7-1/2
445	2-3/8	11	9	8-1/4	6-7/8	7-1/2
445S	2-1/8	11	9	8-1/4	4	7-1/2
445U	2-7/8	11	9	8-1/4	8-3/8	7-1/2
445US	2-1/8	11	9	8-1/4	4	7-1/2
445T	3-3/8	11	9	8-1/4	8-1/4	7-1/2
445TS	2-3/8	11	9	8-1/4	4-1/2	7-1/2
447T	3-3/8	11	9	10	8-1/4	7-1/2
447TS	2-3/8	11	9	10	4-1/2	7-1/2
449T	3-3/8	11	9	12-1/2	8-1/4	7-1/2
503U	3-7/8	12-1/2	10	7	10-3/8	N/A
503US	2-5/8	12-1/2	10	7	5-1/4	N/A
504	2-5/8	12-1/2	10	8	7-5/8	8-1/2
504S	2-1/8	12-1/2	10	8	4	8-1/2
504U	2-7/8	12-1/2	10	8	8-3/8	8-1/2
505	2-7/8	12-1/2	10	9	8-3/8	8-1/2
505S	2-1/8	12-1/2	10	9	4	8-1/2
505U	3-3/8	12-1/2	10	9	10-3/8	8-1/2
505US	2-5/8	12-1/2	10	9	4-3/4	8-1/2
507U	3-3/8	12-1/2	10	11	9-7/8	N/A
507US	2-5/8	12-1/2	10	11	5-1/4	N/A
Large Frame	**U**	**D**	**E**	**F**	**V**	**BA**
5007S	2-1/2	12-1/2	10	11	6-1/2	8-1/2
5007L	3-7/8	12-1/2	10	11	11-1/8	8-1/2
5009S	2-1/2	12-1/2	10	14	6-1/2	8-1/2
50009L	3-7/8	12-1/2	10	14	11-1/8	8-1/2
5011S	2-1/2	12-1/2	10	18	6-1/2	8-1/2
5011L	3-7/8	12-1/2	10	18	11-1/8	8-1/2

ELECTRIC MOTOR SLEEVE BEARING WEAR

The sleeves in sleeve bearing motors are designed to be the component that wears out, rather than the shaft journal itself. One journal will wear out many sets of sleeve bearings before any shaft journal wear is noticeable. Excessive sleeve bearing wear can be suspected if any of the following is noticed:

1. Loss of oil pressure and lubrication occurs, particularly on forced-oil-feed bearings.

2. Excessive vibration occurs because the shaft may climb the bearing wall and then fall back in the seat.

3. The air gap between the rotor and the stator is reduced and the rotor may drag on the stator, or excessive shaft deflections and vibrations may be produced.

While the correct procedure to determine wear of the sleeve bearing involves disassembly of the motor so that the shaft journal outside diameter (OD) as well as the inside diameter (ID) of the sleeve can be measured, it is possible to make a quick check on many motors with a feeler gauge inserted between the top of the shaft journal and the sleeve. The following chart works for either type of inspection.

Electric Motor Sleeve Bearing Wear Chart			
	Factory Supplied Clearance		**Maximum Permitted Wear**
Shaft Size	**Min.**	**Max.**	
0.750″– <1.000″	0.0015″	0.0025″	0.0035″– 0.0040″
1.000″– <1.250″	0.0030″	0.0040″	0.0050″– 0.0060″
1.250″– <2.000″	0.0035″	0.0050″	0.0070″– 0.0080″
2.000″– <2.500″	0.0040″	0.0060″	0.0080″– 0.0090″
2.500″– <3.000″	0.0050″	0.0070″	0.0090″– 0.0105″
3.000″– <4.000″	0.0060″	0.0080″	0.0100″– 0.0115″
4.000″– <5.000″	0.0070″	0.0090″	0.0110″– 0.0125″
5.000″– <6.000″	0.0080″	0.0100″	0.0120″– 0.0140″

INDUSTRIAL MOTOR KEYSEAT DIMENSIONS AND HORSEPOWER RATINGS

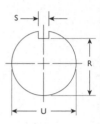

Mechanic note: Industrial keystock is available in standard size, oversize, and undersize. If you are dealing with a repair and the keyseat on the motor or the drive component is worn, an oversize key might be used. Keystock is made from steel, brass, or aluminum, as well as stainless steel, to suit the particular demands of the job at hand.

Shaft	Keyseat Dimensions			
Diameter "U"	R	R (decimal)	S	S (decimal)
3/8	21/64	0.328	FLAT	FLAT
1/2	29/64	0.453	FLAT	FLAT
5/8	33/64	0.516	3/16	0.188
7/8	49.64	0.766	3/16	0.188
1-1/8	63/64	0.984	1/4	0.250
1-3/8	1-13/64	1.203	5/16	0.313
1-5/8	1-13/32	1.406	3/8	0.375
1-7/8	1-19/32	1.594	1/2	0.500
2-1/8	1-27/32	1.844	1/2	0.500
2-3/8	2-1/64	2.016	5/8	0.625
2-1/2	2-3/16	3.188	5/8	0.625
2-7/8	2-29/64	2.453	3/4	0.750
3-3/8	2-7/8	2.875	7/8	0.875
3-7/8	3-5/16	3.313	1	1.000

Standard Horsepower Rating Chart			
1	30	300	1250
1-1/2	40	350	1500
2	50	400	1750
3	60	450	2000
5	75	500	2250
7-1/2	100	600	2500
10	125	700	3000
15	150	800	3500
20	200	900	4000
25	250	1000	———

ADDING A WIRE TO A CONDUIT

Many times additional wires are needed for a new circuit in existing pipe or conduit. They might be phone or Internet cables or speaker wires, or they might be wiring for electrical service. Attempting to push something through the conduit usually wastes time. The solution is to pull the new wire through the pipe. Here is a method that works.

1. Unwind several feet of the light string and feed a little bit of it into one end of the pipe.

2. Start up a shop vac and attach it to the other end of the pipe. Use modeling clay or plumber's putty to seal the connection to produce the most vacuum. The suction will pull the string through the pipe.

3. Watch the end where the string is feeding into the pipe. Make sure it does not get snagged and stop. Be aware of how much string has been pulled in, so you know when to expect the string's exit on the other end.

4. When the length of string sucked into the pipe is a bit more than the total piping run, pull the vacuum hose off the conduit and see if the string is visible.

5. After the light string has been pulled, attach a heavier string or cord to the end and pull this through manually.

6. Once the heavier cord has been pulled through the pipe, attach the final wire and pull this through manually. If there is a need to pull more than one to three smaller wires, connect wires in such a way that the point of connection to the heavy string is not a single clump of wires folded over. This makes for a difficult pull through the conduit. Instead, connect only one wire directly to the string, create one or more loops in this wire 6″ to 8″ down the wire from the connection point. Slip any additional wire(s) several inches into the loop, fold over, and wrap around the first wire.

7. Repeat this as often as required and wrap electrical tape around the string or rope just above the connection point all the way down just beyond the last wrap of the last wire hooked through the loops. The

result should look like the end of a pencil and will pass much more easily through the conduit.

8. Use a metal or fiberglass fish tape or snake. If the conduit run is long, has several bends, has more than 25 percent of its area filled with wires, or the like, the strength and flexibility offered by a fish tape or snake will make pushing into the conduit much easier. Often it can be used to pull new wires directly into the pipe, saving the time of multiple pulls of i string or rope of increasing strength. Fiberglass does not conduct electricity and is highly recommended for use rather than metal fish tapes and snakes.

Conduit Types	
Rigid steel	Heavy-duty pipe with threaded ends. Size is determined by the inside diameter measurement. The conduit is available in different sizes: 1/2, 3/4, 1, 1-1/2, 2, 2-1/2, 3, 3-1/2, 4, 5, and 6.
EMT	Electrical metallic tubing is relatively easy to bend, cut, and form. This light-gauge pipe is also known as a thin-wall conduit. These conduits are available from ½″ to 4″ diameter.
PVC	Used for aboveground and underground applications. This conduit is made of high-quality PVC and is flexible in nature. Conduit is available in different sizes with diameters ranging from 1/2″ to 6″.
GRC	Galvanized rigid conduit is coated with zinc for increasing resistance to corrosion and abrasion.
IMC	Intermediate metal conduit has circular raceways that are rigid and thinner walls as compared to rigid metal conduits.
Nonmetallic conduit	Circular in shape and generally corrugated. This conduit is used in underground installations.

12

CENTRIFUGAL PUMPS

UNDERSTANDING AND TROUBLESHOOTING CENTRIFUGAL PUMPS

Centrifugal pumps are the second-most-used piece of mechanical rotating equipment—second only to an electric motor. The centrifugal pump was invented by Denis Papin, in England, in 1690, but the first practical pump was built and proven by John Appold (another British inventor) in 1851. A centrifugal pump can loosely be described as a device that takes a fluid and "kicks it along." The pump adds energy to the fluid and moves it from one place to another.

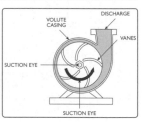

Centrifugal pumps have a rotating component called an *impeller* and a stationary portion called the *housing*. Fluid flows into the inlet eye of the pump, which is flung out by the centrifugal force imparted by the impeller *vanes,* then carried around an increasing spiral cavity called the *volute,* and finally discharged at a higher pressure through the *discharge*.

There is hardly any building, manufacturing site, hospital, process industry, or municipality that does not make use of centrifugal pumps to convey some sort of fluid.

The ability to successfully troubleshoot problems with centrifugal pumps is a great asset to working tradesmen, no matter what their primary skills might be. The following pages offer assistance to craftsmen who might run into pump problems. While the symptoms and causes are widely varied, the troubleshooting checklist should allow the trouble-shooter to resolve and correct the situation.

Symptom	Possible Cause
Pump does not deliver fluid.	Pump is not primed.
	Pump or suction pipe is not completely filled with water.
	Suction lift is too high.
	Suction line contains an air pocket.
	Inlet of suction pipe is insufficiently submerged.
	Suction valve is not open or is only partially open.
	Discharge valve is not open.
	Speed is too low.
	Direction of rotation is wrong.
	Total head of system is higher than design head of pump.
	Parallel operation of pumps is unsuitable for existing conditions.
	Impeller contains foreign matter.
Insufficient capacity is delivered.	Pump or suction pipe is not completely filled with water.
	Suction lift is too high.
	Water contains excessive amount of air or gas.
	Suction line contains an air pocket.
	Air leaks into suction line.
	Air leaks into pump through stuffing box.
	Foot valve is too small.
	Foot valve is partially clogged.
	Inlet of suction pipe is insufficiently submerged.
	Suction valve is only partially open.
	Discharge valve is only partially open.
	Speed is too low.
	Total head of system is higher than design head of pump.
	Parallel operation of pumps is unsuitable for such operation.
	Impeller contains foreign matter.
	Wearing rings are worn.
	Impeller is damaged.
	Casing gasket is defective, permitting internal leakage.

Symptom	Possible Cause
Insufficient pressure has developed.	Water contains excessive amount of air or gas. Speed is too low. Rotation is in wrong direction. Total head of system is higher than design head of pump. Parallel operation of pumps is unsuitable for existing conditions. Wearing rings are worn. Impeller is damaged. Casing gasket is defective, permitting internal leakage.
Pump loses prime after starting.	Pump or suction pipe is not completely filled with water. Suction lift is too high. Water contains excessive amount of air or gas. Suction line contains an air pocket. Air leaks into suction line. Air leaks into pump through stuffing box. Inlet of suction pipe is insufficiently submerged. Water-seal pipe is plugged. Seal cage is improperly located in stuffing box, preventing sealing fluid from entering space to form the seal.
Pump requires excessive power.	Speed is too high. Rotation is in wrong direction. Total head of system is higher than design head of pump. Total head of system is lower than pump design head. Foreign matter in impeller existing conditions There is misalignment. Shaft is bent Rotating part is rubbing on stationary part. Wearing rings are worn. Packing is improperly installed. Type of packing is incorrect for operating conditions. Gland is too tight, resulting in no flow of liquid to lubricate packing.

Symptom	Possible Cause
Stuffing box leaks excessively.	Seal cage is improperly located in stuffing box, preventing sealing fluid from entering space to form the seal. There is misalignment. Shaft is bent. Shaft or shaft sleeves are worn or scored at the packing. Packing is improperly installed. Type of packing is incorrect for operating conditions. Shaft is running off center because of worn bearings or misalignment. Rotor is out of balance, resulting in vibration. Gland is too tight, resulting in no flow of liquid to lubricate packing. Excessive clearance at bottom of stuffing box between shaft and casing is causing packing to be forced into pump interior. Sealing liquid contains dirt or grit, leading to scoring of shaft or shaft sleeve.
Packing has short life.	Water-seal pipe is plugged. Seal cage is improperly located in stuffing box, preventing sealing fluid from entering space to form the seal. There is misalignment. Shaft is bent. Bearings are worn. Shaft or shaft sleeves are worn or scored at the packing. Packing is improperly installed. Type of packing is incorrect for operating conditions. Shaft is running off center because of worn bearings or misalignment. Rotor is out of balance, resulting in vibration. Gland is too tight, resulting in no flow of liquid to lubricate packing. Cooling liquid was not provided to water-cooled stuffing box. Excessive clearance at bottom of stuffing box between shaft and casing is forcing packing into pump interior. Sealing liquid contains dirt or grit, leading to scoring of shaft or shaft sleeve.

Symptom	Possible Cause
Pump vibrates or is noisy.	Pump or suction pipe is not completely filled with water. Suction lift is too high. Foot valve is too small. Foot valve is partially clogged. Inlet of suction pipe is insufficiently submerged. Operation is at very low capacity. Impeller contains foreign matter. There is misalignment. Foundations are not rigid. Shaft is bent. Rotating part is rubbing on stationary part. Bearings are worn. Impeller is damaged. Shaft is running off center because of worn bearings or misalignment. Rotor is out of balance, resulting in vibration. Sealing liquid contains dirt or grit, leading to scoring of shaft or shaft sleeve and resulting noise.
Pump overheats and seizes.	Pump is not primed. Operation is at very low capacity. Parallel operation of pumps is unsuitable for existing conditions. There is misalignment. Rotating part is rubbing on stationary part. Bearings are worn. Shaft is running off center because of worn bearings or misalignment. Rotor is out of balance, resulting in vibration. A mechanical failure inside the pump or failure of the hydraulic balancing device, if any, is causing excessive thrust. Lubrication is inadequate. There is a blocked discharge valve or discharge line. (Caution: Water can turn to steam and cause a steam explosion of the pump case if left unchecked.)

Symptom	Possible Cause
Bearings have short life.	Excessive grease or oil in antifriction-bearing housing, or lack of cooling, is causing excessive bearing temperature. Lubrication is inadequate. Antifriction bearings have been improperly installed (damage during assembly, incorrect assembly of stacked bearings, use of unmatched bearings as a pair, etc.). Shaft seat is too large for proper fit of bearing bore. Dirt is getting into bearings. Bearings are rusting from water getting into housing. Excessive cooling of water-cooled bearing is resulting in condensation of moisture from the atmosphere in the bearing housing. There is misalignment. Shaft is bent. Rotating part is rubbing on stationary part. Rotor is out of balance, resulting in vibration. A mechanical failure inside the pump or failure of the hydraulic balancing device, if any, is causing excessive thrust. Bearings chosen have too little internal clearance for thermal expansion while operating. Bearings chosen for use are not appropriate for the service life required by the pumping situation. There is continued cavitation in the pump housing. Bearing is spinning in housing due to poor fit. Bearing is not flush against shaft shoulder. Bearing was replaced with identical bearing that was removed without checking to make sure this is the correct unit specified by the manufacturer. There is water hammer from fluid. There is excessive vibration damage to bearings from an outside source or transmitted through piping.

PACKING A PUMP

1. Remove all the old packing and thoroughly clean the stuffing box. Check ports and piping to be sure they are not plugged and are free from obstruction. Remove all accumulations inside the stuffing box and on the surface of the shaft.

2. Cut packing rings on an old shaft or mandrel that has the same diameter as the pump shaft. This is very important, as each ring must completely fill the packing space, with no gaps at the ends.

3. Keep the job simple, and cut rings with plain butt joints. There is no advantage to diagonal or skive cuts. The butt joints will be staggered on assembly anyway.

4. Form the first packing ring gently around the shaft, and enter the *ends first* into the box. The installation of this first packing ring is probably the most critical step. The first ring

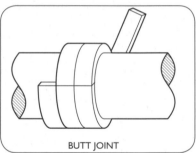

BUTT JOINT

should be gently pushed forward using a tamping tool with a flexible shaft, which aids in keeping the packing square with the shaft as it is being seated. A split bushing of proper size may be used for this job, with good results. In any event, the first ring must be seated firmly against the bottom of the box, with the butt ends together, before additional packing rings are installed.

5. Use a small mirror and flashlight to make sure the ring is seated properly. Never put a few rings into the stuffing box, and try to seat them with the follower. The outside rings will be damaged, and the bottom rings will not be properly seated.

6. Insert additional rings individually, tamping each one firmly into position against the preceding ring. Joints should be staggered to

provide proper sealing and support. If only a few rings are used, the joints should be spaced 120 degrees apart. In installation involving many rings, 90-degree spacing is satisfactory.

7. Position the lantern ring correctly to make sure the front edge is in line with the inlet port. As the packing wears and the follower is tightened, the lantern ring will move forward. When the packing is fully compressed, the lantern ring should still be in line with the inlet port.

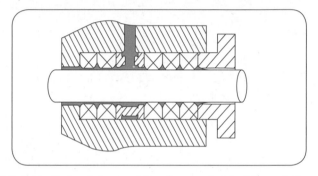

8. Individually insert the remaining rings required to fully pack the stuffing box.

9. Install the gland follower. In a properly designed stuffing box that has been newly packed, the gland follower will enter only a small amount. The gland portion should extend out of the box by an amount to one-third of the packing depth. When the follower is fully tightened, this allows the packing to be compressed to two-thirds of its original volume.

10. Tighten the follower snugly while rotating the shaft by hand. When done in this manner, it is immediately apparent if the follower becomes jammed as a result of cocking or if the packing is overtightened. Slack off and leave finger-tight.

11. Open the valves to allow fluid to enter the equipment. Start the equipment; fluid should leak from the stuffing box. If leakage is excessive, take up slightly on the gland follower. Do not eliminate the leakage entirely; slight leakage is required for satisfactory service. During the

first few hours of run-in operation, the equipment should be checked periodically, as additional adjustment may be required. A leak rate of 45 to 60 drips per minute is considered proper.

Note: About 70 percent of wear is on the outer two packing rings nearest the gland. However, each additional ring does throttle some fluid pressure. On most pumps, there must be enough rings so that if one fails, another does the sealing, and the pump need not be shut down.

CENTRIFUGAL PUMP AFFINITY LAWS EXPLANATION

Centrifugal pumps, being the second-most-used piece of mechanical rotating equipment, as mentioned before, are usually designed to perform properly for a certain system configuration. The pump is expected to deliver a specific flow in gallons per minute (GPM) at a specific head or pressure needed to move the fluid from one location to another.

But as building, municipal, or industrial site needs change, the pump may have to be modified to conform to the new requirements. A larger tank is added to a process, yet the pump is expected to fill the tank in the same amount of time as before. The flow needs to be increased to accomplish this. An existing pump now needs to pump to a higher level of elevation because a structure has been added to a building, increasing the height to which the pump must deliver water to a tank. Or maybe the reverse of this type of situation exists. A decreased demand warrants reducing the flow and saving money by using a smaller-horsepower motor.

There are only two ways of varying head (pressure) or flow from a centrifugal pump. The speed of the pump can be varied

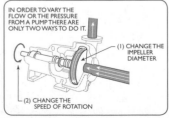

IN ORDER TO VARY THE FLOW OR THE PRESSURE FROM A PUMP THERE ARE ONLY TWO WAYS TO DO IT.

(1) CHANGE THE IMPELLER DIAMETER

(2) CHANGE THE SPEED OF ROTATION

or the size of the impeller can be changed. The pump affinity laws allow calculation for making these changes.

The affinity laws are used to calculate volume capacity, head or power consumption in centrifugal pumps when changing speed– (RPM) or wheel (impeller) diameters.

Law 1: With Impeller Diameter (D) Held Constant

Law 1.1 Flow (GPM) is proportional to shaft speed.

$$\frac{Q_1}{Q_2} = \frac{RPM_1}{RPM_2}$$ *where Q = flow (e.g., GPM) RPM= speed of shaft*

Rearranging the terms provides two very useful versions of this law. If a new flow (Q_2) is required, the following equation calculates the new rotational speed (RPM_2) needed to produce this amount of flow.

Law 1.1a

$$RPM_2 = \frac{Q_2 \times RPM_1}{Q_1}$$

Another common rearrangement solves for a new flow (Q_2) if the speed (RPM_2) is changed.

Law 1.1b

$$Q_2 = \frac{Q_1 \times RPM_2}{RPM_1}$$

Law 1.2 Pressure or head is proportional to the square of shaft speed.

$$\frac{H_1}{H_2} = \left(\frac{RPM_1}{RPM_2} \right)^2 \qquad \begin{array}{l} \textit{where } H = \textit{head or pressure} \\ RPM = \textit{rotational speed of shaft} \end{array}$$

Rearranging the terms provides two very useful versions of this portion of the law. If a new pressure (H_2) is desired, the following equation calculates the new rotational speed (RPM_2) needed to produce this head.

Law 1.2a

$$RPM_2 = \frac{H_2}{H_2} \times RPM_1$$

Another common rearrangement solves for a new head (H_2) or pressure if the speed is changed to RPM_2.

Law 1.2b

$$H_2 = H_1 \times \left(\frac{RPM_2}{RPM_1} \right)^2$$

Law 1.3 Power is proportional to the cube of shaft speed.

$$\frac{P_1}{P_2} = \left(\frac{RPM_1}{RPM_2} \right)^3 \qquad \begin{array}{l} \textit{where } P = \textit{power (horsepower)} \\ RPM = \textit{rotational speed of shaft} \end{array}$$

And once again, a rearrangement of these terms provides a more frequently used version of this portion of the law. If the pump must rotate at a different speed (RPM_2) then the following equation calculates the new power requirement (P_2) that is needed.

Law 1.3a

$$P_2 = P_1 \times \left(\frac{RPM_2}{RPM_1} \right)^3$$

While many modern centrifugal pumping systems use a variable-frequency drive (VFD), which allows the user to vary the speed to change head or flow, an older method of making this type of change involves trimming (machining) the impeller to a smaller diameter. Most pumps carry a nameplate indicating the suction pipe size, the discharge pipe size, and the maximum-diameter impeller that the pump can carry. A nameplate reading "3 × 2—10″" indicates 3″ suction, 2″ discharge, with a 10″ impeller. A second variation of the affinity laws allows new-condition calculations from old ones based on changes to the impeller diameter.

Law 2: With Speed (RPM) Held Constant

Law 2.1 Flow is proportional to impeller diameter (D).

$$\frac{Q_1}{Q_2} = \frac{D_1}{D_2}$$ *where Q = flow (e.g., GPM)*
D = diameter of the impeller

Rearranging the terms provides two very useful versions of this law. If a new flow (Q_2) is required, the following equation calculates the new pump impeller diameter (D_2) needed to produce this amount of flow.

Law 2.1a

$$D_2 = \frac{Q_2 \times D_1}{Q_1}$$

Another common rearrangement solves for a new flow (Q_2) if the impeller diameter (D_2) is changed.

Law 2.1b

$$Q_2 = \frac{Q_1 \times D_2}{D_1}$$

Law 2.2 Pressure or head is proportional to the square of impeller diameter (D).

$$\frac{H_1}{H_1} = \left(\frac{D_1}{D_2}\right)^2 \qquad \begin{array}{l} where\ H = head\ or\ pressure \\ D = diameter\ of\ impeller \end{array}$$

Rearranging the terms provides two very useful versions of this portion of the law. If a new pressure (H_2) is desired, the following equation calculates the new impeller diameter (D_2) needed to produce this head.

Law 2.2a

$$D_2 = \frac{H_2}{H_1} \times D_1$$

Another common rearrangement solves for a new head (H_2) or pressure if the impeller diameter is changed to D_2.

Law 2.2b

$$H_2 = H_1 \times \left(\frac{D_2}{D_1} \right)^2$$

Law 2.3 Power is proportional to the cube of impeller diameter.

$$\frac{P_1}{P_2} = \left(\frac{D_1}{D_2} \right)^3 \quad \begin{array}{l} where\ P = power\ (horsepower) \\ D = diameter\ of\ impeller \end{array}$$

And once again, a rearrangement of these terms provides a more frequently used version of this portion of the law. If the pump is fitted with a different diameter impeller (D_2), then the following equation calculates the new power requirement (P_2) needed.

Law 2.3a

$$P_2 = P_1 \times \left(\frac{D_2}{D_1} \right)^3$$

Note: The affinity laws for pumps work well for the constant-diameter variable-speed case (Law 1) but are less accurate for the case of a constant-speed variable impeller diameter (Law 2)

You have a centrifugal pump that is rated for use with a 10" impeller. Currently the pump is fitted with an 8"-diameter impeller and produces 100 GPM. You need to increase the flow to 125 GPM. What size impeller is needed to accomplish this new flow rate?

Using Equation 2.1a:

$$D_2 = \frac{Q_2 \times D_1}{Q_1}$$

D1 = 8" and flow (Q_1) = 100 GPM. The new flow (Q_2) will be 125 GPM.

$$D_2 = \frac{125 \times 8}{100}$$
$$D_2 = 10"$$

Looks like the maximum impeller diameter of 10" will do the job.

In the preceding example, you calculated that a 10"-diameter impeller would be needed to obtain a new flow of 125 GPM. You checked the original pump configuration using the 8" impeller and found the motor was drawing 30 amps to give 100 GRP. A 40-HP, 3-phase, 440-volt motor is in use, directly coupled to the pump. If you install the new 10"-diameter impeller, will the existing motor be large enough to power the pump? Remember the electrician's rule of thumb that for a 440-volt, 3-phase motor, if you divide the amp draw by 1.25, the result is roughly the amount of horsepower used, so 30 amps divided by 1.25 = a 24-HP draw on the 40-HP motor.

Using Equation 2.3a:

$$P_2 = P_1 \times \left(\frac{D_2}{D_1} \right)^3$$

The existing horsepower draw (P_1) = 24 HP, D_1 = 8″, and D_2 = 10″.

$$P_2 = 24 \times \left(\frac{10}{8}\right)^3$$

$$P_2 = 24 \times (1.25)^3$$

$$P_2 = 46.8 \text{ HP}$$

In short, the 40-HP motor is not large enough; a 50-HP motor is needed to operate the pump under these new conditions.

A centrifugal pump moves storm-sewer water out of a drainage pit to keep your facility from flooding. During some bad storms the pumping rate of 150 GPM is adequate to keep the pit from overflowing. The pump is operated by a VFD motor that allows the RPM to vary. The maximum speed of the motor is 3575 RPM. To produce the flow of 150 GPM, the motor must be operated at 1600 RPM. Storm conditions are brewing and the local radio announcer indicates that this will be the "storm of the century." Based on your last experience with severe weather, you figure that your pump will have to operate at a rate of 300 GPM to handle this one,. What speed does the pump need in order to produce this flow rate?

Using Equation 1.1a:

$$RPM_2 = \frac{Q_2 \times RPM_1}{Q_1}$$

The existing flow (Q_1) = 150 GPM, RPM_1 = 1600, and Q_2 = 300 GPM.

$$RPM_2 = \frac{300 \times 1600}{150}$$

$$RPM_2 = 3200 \text{ RPM}$$

The pump has the RPM to handle the coming storm. If you boost the speed to the maximum of 3575 RPM, how many GPM can you pump? (Hint: Use Law 1.1b.)

335 GPM = GPM

13

INDUSTRIAL FANS

UNDERSTANDING AND TROUBLESHOOTING INDUSTRIAL FANS

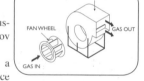

The centrifugal fan was invented by Russian military engineer Alexander Sablukov in 1832.

A centrifugal fan (also called a *squirrel-cage* fan) is a mechanical device for moving air or other gases. It has a fan wheel composed of a number of fan blades, or ribs, mounted around a hub. The hub turns on a driveshaft that passes through the fan housing. The gas enters from the side of the fan wheel, turns 90 degrees, and accelerates due to centrifugal force as it flows over the fan blades and exits the fan housing.

Centrifugal fans can generate pressure increases in the gas stream. Accordingly, they are well suited for industrial processes and air pollution control systems. They are also common in central heating/cooling systems.

The three common blade configurations are forward curved, backward curved, and radial, as show in the following figure.

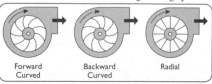

Forward Curved   Backward Curved   Radial

Symptoms	Possible Cause
Impeller is hitting inlet or housing.	Impeller is not centered in inlet or housing. Inlet or housing has been damaged. Impeller is crooked or damaged. Shaft is loose in bearing. Impeller is loose on shaft. Bearing is loose in bearing support. Shaft is bent. Shaft and bearings are misaligned.
Impeller is hitting cutoff.	Cutoff is not secure in housing. Cutoff is damaged. Cutoff is improperly positioned.
Drive is noisy.	Sheave is not tight on shaft (motor or fan). Belts are hitting belt guard. Belts are too loose; adjust for belt stretching after 48 hours of operating. Belts are too tight. Belts have the wrong cross section. Belts are not matched in length on multibelt drive. Variable-pitch sheaves are not adjusted so each groove has the same pitch diameter (multibelt drive). Sheaves are misaligned. Belts are worn. Motor, motor base, or fan is not securely anchored. Belts are oily or dirty. Improper drive was selected. Key is loose.
Coupling is noisy.	Coupling is unbalanced, misaligned, loose, or it may need lubricant. Key is loose. Coupling elastomeric element, grid element, or leaf element is broken and rattling. Coupling is rubbing against safety guard.

Symptoms	Possible Cause
Bearing is noisy.	Bearing is defective. Bearing needs lubrication. Bearing is loose on support. Bearing is loose on shaft. Seals are misaligned. There is foreign material inside bearing. Bearing is worn. There is fretting corrosion between inner race and shaft. Bearing is not sitting on flat surface.
Shaft seal squeals or rubs.	Seal needs lubrication. Seal is misaligned. Shaft is bent. Bearing is loose on support.
Impeller is noisy.	Impeller is loose on shaft. Impeller is defective. Do not run it fast. Contact the service department of the manufacturer. Impeller is unbalanced. Coating is loose. Impeller is worn from abrasive or corrosive material moving through flow passages. Blades are rotating too close to structural member. Blades are coinciding with an equal number of structural members.
Noise is detected in housing.	There is foreign material in housing. Cutoff or other part is loose (rattling during operation).
Drive motor is noisy.	Lead-in cable is not secure. Motor or relay has AC hum. Starting relay chatters. Motor bearings are noisy. Three-phase motor is single phasing. Voltage is low. Motor cooling fan is striking shroud.

Symptoms	Possible Cause
Shaft rattles or rubs.	Shaft is bent. Shaft is undersized; it may cause noise at impeller, bearings, or sheave. Shaft is not centered in fan housing and is touching.
High air-velocity noise is detected.	Ductwork is too small for application. Fan selected is too large for application. Registers or grilles are too small for application. Heating or cooling coil has insufficient face area for application.
Rattle is coming from air stream ducts or continuous whistle is detected,	Dampers are obstructed. Registers are obstructed. Grilles are obstructed. There is an obstruction in sharp elbows or caused by sharp elbows. Ductwork suddenly expands. Ductwork suddenly contracts. Turning vanes are obstructed.
Pulsation or surge noise is detected.	Restricted system causes fan to operate left of peak, indicating change in system or improperly sized fan. Fan is too large for application. Ducts vibrate at same frequency as fan pulsation. There is a rotating stall. There is an inlet vortex surge. Inlet flow is distorted.
Air noise through cracks or holes or when going past obstructions is detected.	Ductwork is leaky. There are fins on the coils, registers, or grills.

Symptoms	Possible Cause
Miscellaneous rattles or rumbles are detected.	Ductwork is vibrating. Cabinet parts are vibrating. Vibrating parts are not isolated from building.
Insufficient air flow is caused by fan.	Impeller is installed backward. Impeller is running backward. Blade angle is set incorrectly. Cutoff is missing or improperly installed. Impeller is not centered with inlet collar(s). Fan speed is too slow. Impeller/inlet is dirty or clogged. Running clearance is incorrect. Inlet cone to wheel fit is incorrect. Inlet vane or damper was improperly set.
Insufficient air flow is being caused by ductwork.	Actual system offers more resistance to flow than expected. Dampers are closed. Registers are closed. Supply ducts are leaky. Insulating duct liner is loose.
Insufficient air flow is being caused by filters.	Filters are dirty or clogged. Replacement filter with greater-than-specified pressure drop is being used.
Insufficient air flow is being caused by coils.	Coils are dirty or clogged. Fin spacing is incorrect.
Insufficient air flow is being caused by recirculation.	Internal cabinet leaks in bulkhead are separating fan outlet (pressure zone) from fan inlets (suction zone). There are leaks around fan outlet at connection through cabinet bulkhead.

Symptoms	Possible Cause
Insufficient air flow is being caused by obstructed fan inlets.	Elbows, cabinet walls, or other obstructions restrict air flow. Inlet obstructions cause more restrictive systems but do not cause increased negative pressure readings near fan inlet(s). Fan speed may be increased to counteract effect of restricted fan inlet(s), but not above factory recommendations for maximum RPM.
Insufficient air flow is being caused by lack of straight duct at fan outlet.	Fans normally used in duct systems are tested with a length of straight duct at fan outlet. If there is no straight duct at the fan outlet, decreased performance may result. If it is not practical to install a straight section of duct at fan outlet, fan speed may be increased to overcome this pressure loss, but not to exceed factory specs for the unit.
Insufficient air flow is being caused by obstruction in high-velocity airstream.	There is an obstruction near fan outlet or inlet. There are sharp elbows near fan outlet or inlet. Turning vanes are improperly designed. There are projections, dampers, or another obstruction in part of system where air velocity is high.
System problems are causing too much air flow.	Ductwork is oversized. Access door is open. Registers or grilles have not been installed. Dampers are set to bypass coils. Filter(s) are not in place. System resistance is low.
There is too much air flow from fan.	Fan speed is too fast. Blade angle is set improperly.

Symptoms	Possible Cause
Static pressure (SP) is wrong.	The velocity pressure at any point of measurement is a function of the velocity of the air and its density. The static pressure at a point of measurement in the system is a function of the system design (resistance to flow), air density, and the amount of air flowing through the system. The static pressure measured in a "loose" or oversized system is less than the static pressure in a "tight" or undersized system for the same air flow rate. In most systems, pressure measurements are indicators of how the installation is operating. These measurements are the result of air flow and thus are useful indicators in defining system characteristics. Field static pressure measurements rarely correspond with laboratory static pressure measurements unless the fan inlet and fan outlet conditions of the installation are exactly the same as the inlet and outlet conditions in the lab.
SP is too low due to change in air density.	Pressures will be lower with high-temperature air or at high altitudes.
Static pressure is low and CFM are high.	System has less resistance to flow than expected. This is a common occurrence. Fan speed may be reduced to obtain desired flow rate. This reduces power (operating cost).
Static pressure is low and CFM are low due to system.	Fan inlet and/or outlet conditions are not the same as tested.

Note: The punkah fan was used in India in 500 BC. It had a canvas-covered frame that was suspended from the ceiling. Servants, known as punkahwallahs, pulled a rope connected to the frame to move the fan back and forth.

Symptoms	Possible Cause
Static pressure is low and CFM are low due to fan.	Impeller is installed backward. Impeller is running backward. Blade angle is set improperly. Cutoff is missing or improperly installed. Impeller is not centered with inlet collar(s). Fan speed is too slow. Impeller/inlet is dirty or clogged. Running clearance is incorrect. Inlet cone to wheel fit is incorrect. Inlet vane or damper is set improperly.
Static pressure is high and CFM are high due to duct.	Actual system is more restrictive (more resistance to flow) than designed. Dampers are closed. Registers are closed. Insulating duct liner is loose.
Static pressure is high and CFM are high due to filters.	Filters are dirty or clogged. Replacement filter has greater-than-specified pressure drop.
Static pressure is high and CFM are high due to coils.	Coils are dirty or clogged. Fins are spaced too closely. Fins have been damaged.
There is high power draw due to fan problems.	Backward-inclined impeller is installed backward. Fan speed is too high. Forward curve or radial blade fan is operating below design pressures. Blade angle is not set properly.

Note: Fans produce air flows with high volume and low pressure, as opposed to compressors, which produce high pressures at a comparatively low volume.

Symptoms	Possible Cause
There is high power usage due to system problems.	Ductwork is oversized. Face and bypass dampers are oriented so coil dampers are open at same time bypass dampers are open. Filter(s) were left out. Access door is open. NOTE: The causes listed for high power usage pertain primarily to radial blade, radial tip, and forward curve centrifugal fans, i.e., fans that exhibit rising horsepower curves. Normally, backward-inclined, backward curve, or backward-inclined airfoil centrifugal fans and axial flow fans do not fall into this category.
There is high power usage due to gas density.	Calculated horsepower requirements are based on light gas (e.g., high temperature), but actual gas is heavy (e.g., cold start-up).
There is high power usage due to improper fan selection.	Fan was not selected at efficient point of rating or is not operating at the best efficiency point (BEP).
Fan fails to operate.	Fuses have blown. Belts are broken. Pulleys are loose. Electricity has been turned off. Impeller is touching housing. Voltage is wrong. Motor is too small and overload protector has broken circuit. Voltage is low, line drop is excessive, or wire size is inadequate. Load inertia is too large for motor. Bearing has seized.

BALANCING A FAN

If a large industrial fan is shaking and vibrating, a good bet is that the unit is unbalanced. While shake in a pump is most likely to be caused by mis-alignment, about 85 percent of cases of fan shake have unbalance as the culprit. Surprisingly, even without sophisticated instruments, a craftsman can balance a fan down a few levels to an acceptable level of shake. It just takes some time, a compass, a ruler, a simple vibration instrument capable of reading in mils (0.001″) or a dial indicator, and some balancing weights. This procedure works if:

■ Unbalance is the main cause of the vibration. A check has been made to eliminate things like bad bearings, misalignment of couplings, bent shafting, and slapping belts as possible sources of vibration. As many factors as possible have been eliminated, and only unbalance remains as a possible cause.

■ The vibration indication is fairly steady. If you are using a meter, the needle must not fluctuate more than ±5 percent of full scale. If using a dial indicator, the needle needs to swing about the same distance from zero each time it moves.

■ The dominant frequency of vibration is equal to the rotating speed of the fan.

It might be good to mention that one of the primary reasons that fans become unbalanced is the presence of dirt and rust. Before attempting any level of corrective action, get a wire brush and thoroughly clean the fan blades. Rap the rust with a ball-peen hammer to shake it loose. If product is stuck to the fan, a power washer might be employed to cut the material from the blades. In many cases, after cleaning, the fan's vibration will be *greatly* reduced.

If you own a vibration instrument, set the amplitude meter to read in mils. If you are using a dial indicator, set the magnet base on a column or support near the fan, but preferably not connected to it. Place the contact point on the bearing housing (not the shaft) closest to the fan and zero the indicator when the fan runs so all the movement is positive. Read the highest value of mils shown by the needle.

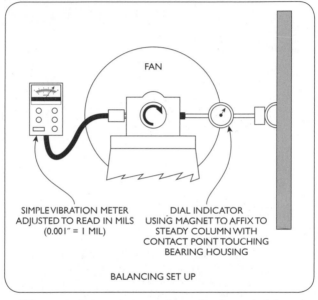

FAN

SIMPLE VIBRATION METER
ADJUSTED TO READ IN MILS
(0.001″ = 1 MIL)

DIAL INDICATOR
USING MAGNET TO AFFIX TO
STEADY COLUMN WITH
CONTACT POINT TOUCHING
BEARING HOUSING

BALANCING SET UP

Either method shown here may be employed to gain information about the unbalance condition of the fan.

In order to balance a fan, a trial weight must be added to the rotor; the changes in vibration readings caused by moving the trial weight allow the mechanic to calculate the correct amount of final weight and the correct position to effectively balance the fan.

The selection of a trial weight depends on the weight of the fan rotor and the speed of the fan. The usual approach is to select a trial weight that can produce an unbalance force at the support bearing equal to 10 percent of the rotor weight supported by the bearing. Here is a worked-out example.

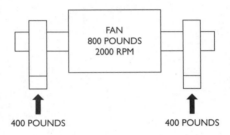

A fan that weighs about 800 lb is supported by two bearings, each supporting 400 lb of rotor weight. The fan turns at 2000 RPM. For this rotor, a suitable trial weight must be selected to produce a force of about 10 percent of the rotor weight, or 40 lbs. (10 percent of 400). The formula for calculating a safe trial weight is given as follows:

$$\text{Force } (lb) = 1.77 \times \left(\frac{RPM}{1000}\right)^2 \times TW(\text{oz-in})$$

Repeating the equation from the previous page and solving for the trial weight (TW):

$$\text{Force } (lb) = 1.77 \times \left(\frac{RPM}{1000} \right)^2 \times Z(\text{oz-in})$$

$$Z\,(\text{oz-in}) = \frac{\text{force } (\text{lb})}{1.77 \times \left(\dfrac{RPM}{1000} \right)^2}$$

Substituting the values from the fan information:

$$Z\,(\text{oz-in}) = \frac{40}{1.77 \times \left(\dfrac{2000}{1000} \right)^2}$$

$$Z\,(\text{oz-in}) = \frac{40}{1.77 \times 22} = \frac{40}{7.08} = 5.65$$

The number of ounce-inches (Z) required would be 5.65. Assume that a weight will be added at a radius of 12″ (if the fan rotor is 24″ in diameter). The actual trial weight is 5.65 oz-in/12″ = 0.47 oz. Rounding off to a value of 1/2 oz gives a practical answer for a trial weight that can be used. Trial weights that are like small C-clamps using setscrews can be purchased from sources accessible on the Internet. In most cases, the C-clamp weights begin at 1/4 oz and increase in size and weight in sensible increments. Values of 1/2, 3/4, 1, 2, 3, 4 oz, and more, are easily obtainable.

The formal procedure for balancing a fan is as follows:

1. Install a suitable method of measuring vibration on the bearing having the highest vibration, usually in the horizontal direction.

2. Measure and record the original amount of unbalance in mils, and record this reading as value "O."

3. Stop the fan, then prepare a trial weight (TW), and install this weight at a position on the fan. It might be easy to position the trial weight at the same distance from the center at which you expect to place the final corrective weight. In addition, placing the weight in line with a key (or keyway) gives a frame of reference for the job. Otherwise, use a china marker or Wite-Out to mark this position.

5. Measure and record the new unbalance vibration as value T1 (for first trial).

6. Stop the fan, then move the trial weight to a position 120 degrees away from the first position at the same radius from the center.

7. Restart the fan and allow it to come up to operating speed.

8. Measure and record the new unbalance vibration as value T2 (for second trial).

9. Stop the fan again and move the trial weight to a new position 120 degrees away from the T2 position at the same radius from the center.

10. Measure and record the new unbalance vibration as value T3 (for the third trial).

11. Construct a diagram as shown here by using the O, T1, T2, and T3 values obtained in the trial runs of the fan. A scale of 1/8″ = 1 mil of unbalance is used in this drawing. Let us imagine that O = 10 mils, T1 = 15 mils, T2 = 12 mils, and T3 = 9 mils.

12. At a scale of 1/8″ = 1 mil, then (divide by 8):

    O = 10 mils = 1-1/4″
    T1 = 21 mils = 2-5/8″
    T2 = 12 mils = 1-1/2″
    T3 = 9 mils = 1-1/8″

13. Use a compass set at a radius of O or 1-1/4″ and draw the O circle.

14. Choose a point at the 12 o'clock or 0-degree position, place the compass point there, and scribe an arc on the circle. Remember that a compass set to the same radius as the circle can be used to scribe around the circumference and will divide the circumference into six equal parts. Move the compass point to the place where the arc and circle meet and use that as the center for a new arc. Each mark indicates 60 degrees, so every two marks is 120 degrees. This is an easy way to divide the circumference into 120-degree increments to duplicate the places you moved the weight. You should have marks at the 0-degree, 120-degree, and 240-degree places around the circumference of the O circle.

Imagine the 0-degree position to be in line with the keyway or the location on the fan at which you placed the trial weight for the first trial, or T1. Then you moved the weight 120 degrees for the second trial, or T2, and finally you made another 120-degree move to obtain the reading for the third trial, or T3. The last position is 240 degrees away from the original placement of the weight for T1.

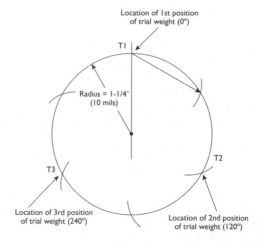

Location of 1st position
of trial weight (0°)

T1

Radius = 1-1/4″
(10 mils)

T2

T3

Location of 3rd position
of trial weight (240°)

Location of 2nd position
of trial weight (120°)

15. Now reset the compass for a radius equal to the amount of unbalance when you ran T1. That value was 21 mils, or a radius of 2-5/8″ (1/8″ = 1 mil).

16. Place the compass point at the T1 intersection on the O circle and scribe an arc.

17. Now reset the compass for a radius equal to the T2 value of unbalance: 12 mils, or 1-1/2″.

18. Place the compass point at the T2 intersection on the O circle and scribe an arc.

19. Reset the compass for a radius equal to the amount of unbalance when you ran T3. That value was 9 mils, or a radius of 1-1/8″ (1/8″ = 1 mil).

20. Place the compass point at the T3 intersection on the O circle and scribe an arc.

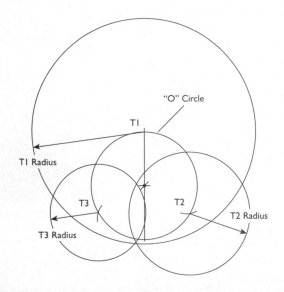

The final figure should look like the one shown previously. An original O circle divided into 120 increments around the circumference, with each of the increment points used as a center for drawing (three) arcs where the radius equals the respective unbalance during each trial run.

21. Notice there is a point where all three arcs meet. Draw a straight line from the middle of the O circle to this meeting point. See the *heavy* black line in the diagram shown here.

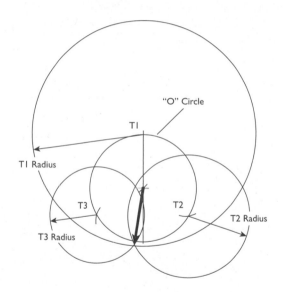

22. Carefully measure the length of line. For clarity we will call it the T arrow. In this case T = 1-3/8″ so, using the scale of 1/8″ = 1 mil (as before), T = 11 mils.

23. You are ready to calculate the correct weight (CW) needed to balance the fan.

$$CW = TW \times \left(\frac{O}{T}\right)$$

The CW is equal to the value of the trial weight times the original amount of vibration in mils (O) divided by the value of the T arrow in mils as measured and scaled from the drawing (1/8″ = 1 mil).

$$CW = 0.5 \times \left(\frac{10 \text{ mils}}{11 \text{ mils}}\right)$$

$$CW = 0.45 \text{ oz}$$

24. The weight will be placed on the fan in the angular position shown by the T arrow. Notice that the T arrow is a little more than 180 degrees from the original placement of the trial weight at T1. Remember that you were in line with the keyway; that will help you locate the position for the placement of the correct weight to reduce the amount of shake due to unbalance.

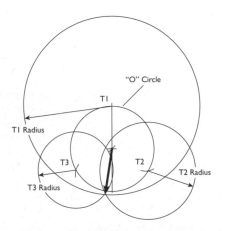

25. The *exact* position for placement of the CW would be located on the rotor by measuring the angle between the 0-degree position on the diagram and the T arrow. This can be accomplished using a protractor.

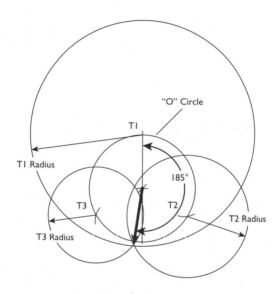

26. Accurate measurement shows the correct placement to be 185 degrees from the placement of the trial weight at the 0-degree position (T1). It's easy if you also made T1 line up with the keyway on the shaft.

14

WELDING

OXYACETYLENE GAS WELDING

Acetylene supply cylinders are usually painted black and have cylinder

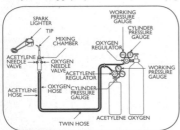

valves. The cylinders contain a porous filler material wetted with acetone that allows the acetylene to safely be contained in the cylinder at 250 psig. Always use an acetylene cylinder in the upright position to keep any acetone from being drawn out of the tank. Open the cylinder valve only 1 to 1-1/2 turns, leaving the valve wrench on the valve in the event that it has to be shut off quickly. Acetylene should never be used at a pressure that exceeds 15 psig, as it becomes highly unstable, which, depending on the condition, could cause it to decompose and explode. As with the oxygen cylinder, make sure the cylinder valve is clean before installing the regulator.

Lowering the bottle pressure of 2200 psig (oxygen) and 250 psig (acetylene) to a desired working pressure for use in the torch is accomplished by using an adjustable pressure-reducing regulator. The regulator also maintains a steady working pressure as the cylinder pressure drops from use. Regulators work by admitting the high cylinder pressure through a valve that is operated by a flexible diaphragm.

When burned alone, acetylene can produce a flame temperature of about 4000°F. With the addition of oxygen, a flame temperature in excess of 6000°F can be achieved, making acetylene ideal for welding and cutting. An oxyacetylene outfit is portable, less expensive, and more versatile than an electric welding setup. By using the proper tips, rods, and fluxes, almost any metal can be welded, heated, or cut using the oxyacetylene process.

Oxygen supply cylinders are usually painted green and can hold pressures as high as 2200 psig. All cylinders have valves and are fitted with a screw-on steel cap that protects the valve during storage or shipping. If oxygen comes into contact with oil or grease, it will burst into flame. Never use oil or grease on oxygen cylinder valves or regulators. Make sure hands and gloves are free

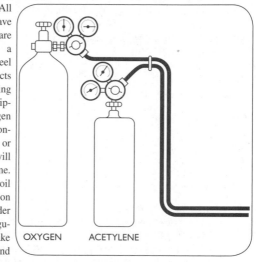

of oil and grease before handling cylinders. Crack open the cylinder valve, then close it before installing the regulator to clear the valve of any dirt. With the regulator installed, always crack the cylinder valve open first, then open it fully. This lessens the chance of recompression, which is caused by high cylinder pressure entering the regulator and heating up and damaging it.

Turning the regulator adjusting knob in or out causes a spring in the regulator to operate the diaphragm, which opens or closes a valve in the regulator. This in turn regulates the outlet pressure and flow. By turning the adjusting knob in, the flow and pressure increase; turning it out decreases the flow and pressure. Most regulators have two gauges. One shows the inlet pressure from the cylinder (the high-pressure gauge), and the other (the low-pressure gauge) shows the working pressure being supplied from the regulator. Oxygen regulators have right-hand threads, and regulators for acetylene have left-hand threads.

There is a groove around fuel connections that indicates a left-handed thread. All outlet (low-pressure) gauges on acetylene regulators have their gauge scales marked in red, starting at 15 psig.

Double-line rubber hoses connect the cylinder regulators and torch. The oxygen line is green; the acetylene line is red.

The welding tip is mounted on the end of the torch handle and, through it, the oxygen-and-fuel gas mixture feeds the flame. Tips are available in a variety of shapes and sizes to fit most any welding job and are identified by number. The larger the number, the larger the hole in the tip and the thicker the metal that can be welded or cut. Welding tips have one hole, and cutting tips have a centrally located hole with a number of smaller holes located around it in a circular pattern. The cutting oxygen comes from the center hole with the preheat flame coming from the holes around it. Many factors determine the tip size to use, but mainly the thickness of the metal to be welded or cut determines which tip size to use.

The basic method to cut with a torch is as follows:

- Open the gas valves to the correct pressure. (A rule of thumb is 40 psi for oxygen and 8 psi for acetylene.)
- Ignite the gas with a striker.
- Set the flames to the correct length for the type of work you're doing. (There are basically three kinds of flame: oxidizing flame, neutral flame, and carburizing flame.)
- Preheat the steel.
- Press the oxygen valve to begin the cut.
- Advance the tip smoothly through the cut.

The reason this works is that the torch is oxidizing the metal before its melting point, and the oxide produced then also melts at a lower temperature than the parent metal and is blown away by the high-pressure oxygen stream. It's almost like magic!

Cutting Torch Tip Selection Chart		
Thickness of Steel Plate (in.)	**Diameter of Cutting Tip Orifices (in.)**	**Approximate Cutting Speed (in/min)**
1/8	0.020–0.040	16–32
1/4	0.030–0.060	16–26
3/8	0.030–0.060	15–24
1/2	0.040–0.060	12–23
3/4	0.045–0.060	12–21
1	0.045–0.060	9–18
1-1/2	0.060–0.080	6–14
2	0.060–0.080	6–13
3	0.065–0.085	4-11
4	0.080–0.090	4–10
5	0.080–0.095	4–8
6	0.095–0.105	3–7
8	0.095–0.110	3–5
10	0.095–0.110	2–4
12	0.110–0.130	2–4

15

MISCELLANEOUS INFORMATION

RULES OF THUMB AND TRADE TRICKS

No shim available?

A Post-it Note measures 0.004″.

A credit card measures 0.030″ (roughly 1/32″).

No feeler gauge?

Beverage-can lid: 0.014″ (typical ignition-point gap).

Jumbo paper clip: 0.040″.

No scale handy?

1 gal water = 8.4 lb.

1 gal gasoline = 6.2 lb.

1 gal diesel fuel = 7.2 lb.

Need something to measure length?

A quarter measures 0.955″ in diameter (1″).

Standard credit card = 3-3/8″ × 2-1/8″.

A 1″ box wrench is roughly 12″ long.

Need to measure a liquid?

A 12-oz soda can contains 1.5 cups.

A 1-qt oil container cap holds about 1 oz.

A spray-paint cap holds 4.5 oz.

Rough guess on motor amperage draw?

At 575 volts, a 3-phase motor draws 1 amp per HP.

At 460 volts, a 3-phase motor draws 1.25 amps per HP.

At 230 volts, a 3-phase motor draws 2.5 amps per HP.

At 230 volts, a single-phase motor draws 5 amps per HP.

At 115 volts, a single-phase motor draws 10 amps per HP.

Don't have a temperature conversion chart?

°F to °C: Subtract 30 and take 1/2 of what is left.

70°F is about (70 − 30 = 40, then 1/2 of 40) = 20°C.F

(Actual correct answer is 21°C.)

° to °F: Double and add 30.

40°C is about (40 × 2 = 80, then add 30) = 110°F.

(Actual correct answer is 104°C.)

What is 55°C in °F? (Answer: About 140°F; actual: 131°F.)

Doing a precision alignment?

No more than four (4) shims to be used under any foot. Replace multiple thin shims with single-thickness plates.

Measurement:

A U.S. dollar bill is exactly 6″ wide. This comes in handy if you need to measure something accurately and you don't have a tape measure. Two widths equal a foot exactly. A dollar bill folded in half becomes exactly 3″ wide.

Plumbing:

Every 2′ of height in a building drops the water pressure by about 1 psi.

How long does a light last?

- Fluorescent T8: 20,000 hours, or 3 years at 18 hours/day
- Fluorescent high-output T5: 20,000 hours, or 3 years at 18 hours/day
- Metal halide: 20,000 hours, or 3 years at 18 hours/day
- Incandescent A lamp: 750 hours, or 2 months at 18 hours/day
- Incandescent MR16 halogen: 10,000 hours (with special lamp), or 1-1/2 years at 18 hours/day
- LED: 100,000 hours, or +15 years at 18 hours/day

Pipe sizes?

Steel, aluminum, and PVC pipes always have the same outside diameter (OD) regardless of wall thickness/schedule. A pipe has a larger OD than its rated size. A tube always has the same OD regardless of wall thickness. Inner diameter (ID) = OD minus 2 times the wall thickness.

Telescoping PVC pipe?

Any 1.5″ schedule 40/80 will fit into 2″ schedule 80 pipe.

Any 2.5″ schedule 40/80 will fit into 3″ schedule 80 pipe.

Can't remember Ohm's law?

Twinkle twinkle little star $E = I \times R$.

Need to measure longer distances without a tape?

Calibrate yourself. Mark off 100′ with a tape measure. Now pace it off and count your strides. Do it a few times until you are sure. Next time you need to roughly measure 200′, it should be easy.

Wiring a standard receptacle?

Black wire to brass (B to B).

Ground wire to green (G to G).

White is what's left, it must go to silver.

The ground connection on the face of the receptacle is toward the ground (at the bottom, toward the earth).

Trying to judge a motor or bearing temperature by touch?

Feels cool = 77°F to 95°F

Neutral = 98.6°F (human body temperature)

Warm = 104°F

Warmer (you can hold your finger on it) = 122°F

Very warm = 140°F (can be tolerated for about 2 sec)

Hot = 158°F (can be tolerated for about 1 sec)

Makes a blister = 200°F or more

Spit on the surface and it boils = 212°F or higher

Accurate checking of your tire pressure?

For every 10°F change in air temperature, the tire's inflation pressure changes by about 1 psi (up with higher temperatures and down with lower temperatures). Pressure listed on tires is correct for temperature of 70°F.

Need a 1-lb weight?

A pint's a pound the whole world round!!!

1 pint of water weighs almost exactly 1 lb.

Rigging safety:

The safe working load of wire rope is the square of the rope diameter multiplied by 8, in tons. The diameter is measured with a caliper from the very outer point of one strand to the outer point of the opposite strand, *not* so that either jaw of the caliper is flush with more than one strand simultaneously.

Cutting threads with a die:

Turn on 1 turn, back 1/2 turn.

Sizing an electric motor:

If a piece of machinery can be operated by hand, a 1/4-HP motor is sufficient. If a gasoline engine is to be replaced with an electric motor, an electric motor approximately 2/3 the horsepower rating of the engine is adequate.

Removing a broken lightbulb:

If you have to change a lightbulb whose glass is broken, you can press a potato into the metal base to unscrew the remains of the bulb from the fixture.

Choice of a key?

Use square keys for shafts less than 6.5″ in diameter and rectangular keys if greater.

Fastening:

The length of the bolt head is equal to the diameter of the screw. The minimum thread engagement is equal to the diameter of the screw, but 1.5 × the diameter is better.

Figuring the diameter of a screw thread:

The #0 machine screw diameter is 0.060″; add 0.013″ for each # above this (e.g., the diameter of a #8 screw is 0.060 + (8 × 0.013) = 0.164″).

Hand tapping:

Turn in 1/4 to 1/2 turn, then out 3/4 turn.

Removing stickers from paper?

(This applies to things like price stickers.) Douse with lighter fluid (naphtha) and let soak. Slowly lift sticker. Paper will not be harmed. Works well on old books and collector's documents.

Welding hood:

Put Rain-X on your outside clear lens, which helps keep it clean and is very easy to wipe off, with fewer changes needed.

DIVIDING A DIMENSION IN HALF

Almost every trade requires the ability to measure a surface and find the half mark. Locating the center of a room for laying tile, finding the middle of a piece of wood to divide into two equal pieces, finding the center of a wall to hang a painting are examples that come to mind. Finding half of a dimension is easy if the original measurement is a whole number, such as 4, or 20, or 17. For the most part, people accustomed to making measurements handle that type of division in their heads, ending up with 2, or 10, or 8-1/2, respectively.

It's when the number also contains a fraction that the trouble seems to begin. A whole number and a fraction divided in half is easy when the whole number is even. Here's the rule for doing it in your head.

<u>Rule for dividing an even number and a fraction in half:</u>

■ Divide the whole number by 2.
■ Leave the numerator (top) of the fraction alone as the new numerator.
■ Double the denominator (bottom of the fraction as the new denominator.

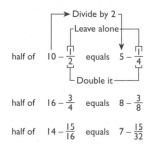

$$\text{half of} \quad 16 - \frac{3}{4} \quad \text{equals} \quad 8 - \frac{3}{8}$$

$$\text{half of} \quad 14 - \frac{15}{16} \quad \text{equals} \quad 7 - \frac{15}{32}$$

You get the idea. Can you do it in your head?

The problem with doing it in your head seems to be when you need to divide an odd number and a fraction in half. That's not so easy. But if you know the rule, it is! Here's the rule for doing it in your head.

<u>Rule for dividing an odd number and a fraction in half:</u>

■ Divide the odd whole number by 2 to get the biggest whole number, and forget the rest
■ Add the numerator and denominator of the existing fraction to get the numerator of the result.
■ Double the denominator of the existing fraction to get the denominator of the result.

A couple of examples will help you do it in your head, too. Just follow the rules.

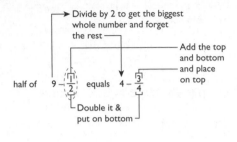

$$\text{half of} \quad 17 - \frac{3}{8} \quad \text{equals} \quad 8 - \frac{11}{16}$$

$$\text{half of} \quad 27 - \frac{7}{16} \quad \text{equals} \quad 13 - \frac{23}{32}$$

Try a few on paper until you get fast at it. You'll be surprised how many times you'll use this technique on the job.

SIMPLE ROPE LOAD CALCULATIONS

One of the most frequently used rigging implements is good old-fashioned rope. Because no rigger can be expected to remember the safe working loads of ropes, rules of thumb can be used to estimate the loads. The following rules of thumb work well for *new* ropes when load tables are not available. SWL = safe working load.

<u>Manila Rope</u>

Change the rope diameter into eighths of an inch.

Square the numerator and multiply by 20.

Here are three examples:

$$1/2'' \text{ manila rope} = 4/8'' \text{ diameter}$$

$$SWL = 4 \times 4 \times 20 = 320 \text{ lb}$$

$$5/8'' \text{ manila rope}$$

$$SWL = 5 \times 5 \times 20 = 500 \text{ lb}$$

$$1'' \text{ manila rope} = 8/8'' \text{ diameter}$$

$$SWL = 8 \times 8 \times 20 = 1280 \text{ lb}$$

<u>Nylon Rope</u>

Change the rope diameter into eighths of an inch.

Square the numerator and multiply by 60.

<u>Polypropylene Rope</u>

Change the rope diameter into eighths of an inch.

Square the numerator and multiply by 40.

<u>Polyester Rope</u>

Change the rope diameter into eighths of an inch.

Square the numerator and multiply by 60.

<u>Polyethylene Rope</u>

Change the rope diameter into eighths of an inch.

Square the numerator and multiply by 35.

What is the safe working load that a 5/8″ nylon rope can handle?
Answer: $5 \times 5 = 25 \times 60 = 1500$ lb.

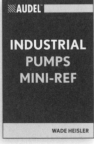